Contents

The content listed in bold is only specified to be assessed at A-level.

Introduction 5

1 Arithmetic and numerical computation

Appropriate units in calculations 7

Expressions in decimal and ordinary form 20

Ratios, fractions and percentages.................................27

Estimating results 33

2 Handling data

Significant figures...36

Arithmetic mean 42

Identifying uncertainties in measurements.............................46

3 Algebra

Understanding symbols 50

Changing the subject of an equation 50

Solving algebraic equations...53

Using calculators to find logs...59

4 Graphs

Plotting graphs...64

The slope and intercept of a linear graph..................................68

Calculating the rate of change 73

Tangents and measuring the rate of change 76

5 Geometry and trigonometry

Representing shapes of molecules in 2D and 3D form................. 85

Exam-style questions

Worked examples... 91

AS and A-level questions... 97

A-level only questions.. 100

Appendix

Appendix... 102

Essential Maths Skills
for AS/A-level
Chemistry

Nora Henry

HODDER
EDUCATION
AN HACHETTE UK COMPANY

Hodder Education, an Hachette Company, Carmelite House, 50 Victoria Embankment, London, EC4Y 0DZ

Orders
Hachette UK Distribution, Hely Hutchinson Centre, Milton Road, Didcot, Oxfordshire, OX11 7HH
tel: 01235 827827
e-mail: education@hachette.co.uk

Lines are open 9.00 a.m.–5.00 p.m., Monday to Friday. You can also order through the Hodder Education website: www.hoddereducation.co.uk

ISBN 978-1-4718-6349-3

First printed 2016
Impression number 5
Year 2022

Typeset in India

Cover illustration: Barking Dog Art

Printed and bound by CPI Group (UK) Ltd, Croydon, CR0 4YY

Hachette UK's policy is to use papers that are natural, renewable and recyclable products and made from wood grown in well-managed forests and other controlled sources. The logging and manufacturing processes are expected to conform to the environmental regulations of the country of origin.

Introduction

Tapputi-Belatikallim is considered to be the world's first chemist. She was a perfume maker in 2000 BC who distilled flowers, oil, myrrh and other aromatic materials to make perfume. As a chemist, she was continually changing quantities and techniques to develop her product. Today in the twenty-first century chemists continue to experiment as they develop new fibres, drugs, foods and cosmetics, create new processes to reduce future energy use and determine the composition of different substances. All chemists need to process and analyse the data obtained from experimental work. This involves carrying out calculations, using equations and plotting graphs — all of which are elements of maths. Maths is central to the study of chemistry and to develop your skills, knowledge and understanding in chemistry, you must also develop your maths skills.

Overall, at least 20% of the marks in assessments for AS and A-level chemistry will require the use of maths skills. These skills will be applied in the context of chemistry, for example, in calculating the concentration of a solution, determining an entropy change or finding the gradient of a graph to determine the order of a reaction. The standard of maths required is that of level 2 or above, and this book will take you through the different skills at the depth needed.

The Department of Education has published a specification giving the maths requirements for GCE chemistry for all examination boards, and hence no matter what examination board specification you are following, the same maths skills are required. This specification is shown in detail on pages 102–104.

Perhaps you struggle with maths, or are not confident with formulae, equations and numbers. This book starts with simple skills and progresses step by step through the maths concepts required, enabling you to develop a sound mathematical foundation, which will in turn benefit your chemical proficiency.

Questions which involve maths often include the command words shown in the table below. A command word is the verb in the question which the examiner uses to tell you what to do and it is useful to understand those shown.

Command word	Meaning
calculate	Obtain a numerical answer, showing relevant working. If the answer has a unit, this must be included.
determine	Use given data or information to obtain an answer which must be quantitative, or must show how the answer can be reached quantitatively.
estimate	Assign an approximate value.
plot	Produce a graph by marking points accurately on a grid from data provided and then drawing a line of best fit through these points. A suitable scale and appropriately labelled axis must be included.

How to use this book

This book should be used in conjunction with your textbook and notes to help you practise and consolidate maths concepts which you may find difficult, and which may hamper your progress in chemistry.

When beginning a new topic in your course it is best to first study the appropriate maths skills required. For example, when studying amount of substance calculations, check that you understand the concept of rearranging the formula of an equation and substituting values into an equation by working through the appropriate sections, or when starting pH calculations, first work through the section in this book on logarithms.

Each topic starts with a detailed explanation of the mathematical skill, followed by worked examples. Guided questions are then included. When working through these it is important that you work actively and use your calculator and paper and pen to complete the questions. Finally, to further consolidate the skill there are practice questions which you should complete and mark using the online answers — these give detailed explanations of the solutions. All questions are written in a chemistry context and this will further build on your understanding of chemistry. When you have a sound knowledge of the maths concepts, you should find the chemistry topic much easier to understand. Exam-style chemistry questions are found at the end of the book and these could be used for final topic or module consolidation or for revision work.

Playing a musical instrument well takes a lot of practice. Few musicians are immediately skilful, but over time their expertise develops. Learning a new musical piece takes effort and repeated rehearsing in order to improve. Preparing for the maths-type questions for A-level chemistry is a similar process. Last minute revision will not be effective. You need to master the skills of maths as you progress through your course. Carefully working through each skill area in this book and completing the practice questions will help you consolidate your maths skills and lead to a better understanding of chemistry.

Full worked solutions to the guided and practice questions and exam-style questions can be found online at www.hoddereducation.co.uk/essentialmathsanswers.

1 Arithmetic and numerical computation

Appropriate units in calculations

Laboratory work is central to the study of chemistry, and throughout your course there will be many opportunities for you to use practical work to investigate, record and process data. Some measurements recorded in experimental work may be **qualitative** and do not include a numerical value. Observations such as 'bubbles' or 'the colourless limewater changed to cloudy' are qualitative measurements. **Quantitative** measurements, such as mass, temperature and volume, do include a numerical value. It is essential that units are included for all quantitative measurements. For example, stating that the mass of a portion of a solid is 0.4 says very little about the actual mass of the solid — is it 0.4 g or 0.4 kg? The units are essential.

Some of the quantitative measurements and the units you may use are shown in Table 1.1.

Table 1.1

Quantitative measurement	Units
temperature	degrees Celsius (°C) or kelvin (K)
volume	centimetre cubed (cm³), decimetre cubed (dm³), metre cubed (m³)
mass	milligrams (mg), grams (g), kilograms (kg), tonnes (t)
pressure	pascals (Pa) or kilopascals (kPa)

Compound units are those composed of more than one unit. For example:
- $mol\,dm^{-3}$ which is the unit of concentration
- $g\,cm^{-3}$ which is the unit of density
- $kJ\,mol^{-1}$ which is the unit of energy.

Converting between units — temperature

The kelvin temperature scale includes the value of absolute zero, 0 K, which is the value at which the movement of all particles stops. This temperature is –273 °C. Temperatures in kelvin are either zero or a positive number. When writing a temperature in kelvin there is no need for a degree sign, the unit is represented simply as K. To convert temperatures from degrees Celsius (°C) to kelvin (K), add 273. To change from kelvin (K) to degrees Celsius (°C), subtract 273.

Figure 1.1 Conversion between °C and K

Worked examples

a   **Convert 35 °C to kelvin.**

To convert from °C to kelvin add 273.

$35 °C + 273 = 308 K$

b   **Convert −11 °C to kelvin.**

To convert from °C to kelvin add 273.

$-11 °C + 273 = 262 K$

 Guided question

Copy out the workings and complete the answers on a separate piece of paper.

1   **Convert 203 K to °C.**

To convert from kelvin to °C subtract 273.

$203 K - \underline{\hspace{1cm}} = \underline{\hspace{1cm}}$

> **TIP**
>
> Note that when converting from K to °C a negative number may be obtained, but when converting from °C to K the number will always be positive.

 Practice question

2   Carry out the following unit conversions.
    a   246 K to °C
    b   −35 °C to K
    c   344 K to °C
    d   −19 °C to K
    e   45 K to °C
    f   192 K to °C

Converting between other units

Volume

Volume is usually measured in centimetre cubed (cm^3), decimetre cubed (dm^3) or metre cubed (m^3). You need to remember that:

■ $1000 \, cm^3 = 1 \, dm^3$
■ $1000 \, dm^3 = 1 \, m^3$

Figure 1.2 shows how to convert between the different volume units.

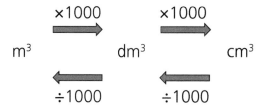

Figure 1.2 Converting between volume units

Mass

Mass can be measured in milligrams (mg), grams (g), kilograms (kg) and tonnes (t). You need to remember that:

- 1 t = 1000 kg
- 1 kg = 1000 g
- 1 g = 1000 mg

Figure 1.3 shows how to convert between mass units.

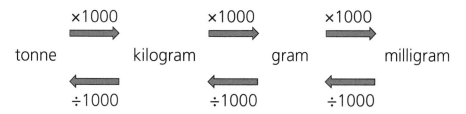

Figure 1.3 Converting between mass units

Pressure

There are many different units of pressure, but the one most often used in chemistry is the pascal. The symbol for a pascal is Pa. You need to remember that:

- 1 kilopascal = 1000 pascal

Figure 1.4 shows how to convert between pressure units.

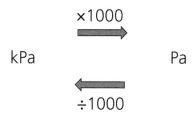

Figure 1.4 Converting between pressure units

A Worked examples

a   **What is 35 cm³ in dm³?**

To convert from cm³ to dm³ divide by 1000.

$$\frac{35 \, cm^3}{1000} = 0.035 \, dm^3$$

b   **What is 0.02335 m³ in dm³?**

To convert from m³ to dm³ multiply by 1000.

$$0.02335 \, m^3 \times 1000 = 23.35 \, dm^3$$

c   **What is 2.2 tonnes in grams?**

Two conversions are needed here:

tonne $\longrightarrow$ kilogram $\longrightarrow$ gram

Step 1: to convert from tonnes to kilograms multiply by 1000.

$$2.2 \, t \times 1000 = 2200 \, kg$$

Step 2: to convert from kilograms to grams multiply by 1000.

$$2200 \, kg \times 1000 = 2\,200\,000 \, g$$

d   **Convert 1.2 kPa into Pa.**

To convert from kPa to Pa multiply by 1000.

$$1.2 \, kPa \times 1000 = 1200 \, Pa$$

B Guided questions

Copy out the workings and complete the answers on a separate piece of paper.

1   **What is 12 000 cm³ in m³?**

Two conversions are needed here:

cm³ $\longrightarrow$ dm³ $\longrightarrow$ m³

Step 1: to convert from cm³ to dm³ divide by 1000.

$$\frac{12\,000 \, cm^3}{1000} = \underline{\hspace{2cm}} dm^3$$

Step 2: to convert from dm³ to m³ divide by 1000.

$$\frac{\underline{\hspace{1cm}}}{1000} = \underline{\hspace{2cm}} m^3$$

2   **What is 1.2 kg in g?**

To convert from kg to g multiply by 1000.

$$1.2 \, kg \times 1000 = \underline{\hspace{2cm}} g$$

Remember that a kilogram is a larger unit than a gram so you would expect to get a bigger number of grams.

> **TIP**
>
> Always think about your answer: a cm³ is a smaller unit than a dm³ so you would expect to get a smaller number of dm³.

 **Practice question**

3 Carry out the following unit conversions.

   a 1.2 dm³ to cm³

   b 420 cm³ to dm³

   c 3452 cm³ to m³

   d 1.4 t to g

   e 4 kg to g

   f 101 kPa to Pa

Calculations which often involve conversion of units

There are several topic areas in which you may often need to convert units, including:

- amount of substance, where you may need to calculate moles using the expressions

$$\text{amount (in moles)} = \frac{\text{mass (g)}}{M_r} \qquad \text{or} \qquad \text{amount (in moles)} = \frac{\text{mass (g)}}{A_r}$$

   M_r = relative molecular mass A_r = relative atomic mass

- ideal gas equation
- entropy.

> **TIP**
>
> Make sure you are familiar with using your Periodic Table to find relative atomic masses (A_r).

Amount of substance

 Worked example

Calculate the amount, in moles, present in 2.4 tonnes of magnesium.

- To calculate the amount in moles use the expression

$$\text{amount (in moles)} = \frac{\text{mass (g)}}{A_r}$$

- Before using this expression, the mass of magnesium must be converted from tonnes to grams.

   tonnes $\xrightarrow[\times 1000]{}$ kilograms $\xrightarrow[\times 1000]{}$ grams

- A mass in tonnes is often given. It is useful to remember that to convert from tonnes to grams you can multiply by 10^6.

Step 1: mass of magnesium in grams = 2.4 t × 1000 × 1000 = 2 400 000 g

Step 2: amount (in moles) = $\dfrac{\text{mass (g)}}{A_r}$

$$= \frac{2\,400\,000\,\text{g}}{24.3} = 98\,765.4\,\text{mol}$$

TIP

Some examination boards may ask you to calculate the number of moles rather than the amount in moles — this is answered in the same way.

B Guided question

Copy out the workings and complete the answers on a separate piece of paper.

1 **Calculate the amount, in moles, present in 3.312 kg of lead nitrate, $Pb(NO_3)_2$.**

To calculate the amount, in moles, use the expression

$$\text{amount (in moles)} = \frac{\text{mass (g)}}{M_r}$$

Step 1: convert the mass from kg to g by multiplying by 1000.

3.312 kg × 1000 = _____

Step 2: find the M_r of lead nitrate by adding the individual relative atomic masses — there is 1 Pb atom, 2 N atoms and 6 O atoms.

Step 3: substitute the mass in grams and the M_r into the equation to calculate your final answer.

C Practice questions

2 Calculate the amount, in moles, present in 17 kg of ammonia (NH_3).

3 Calculate the amount, in moles, present in 2.1 tonnes of iron(III) oxide (Fe_2O_3).

4 Calculate the amount, in moles, present in 2.42 kg of magnesium nitrate ($Mg(NO_3)_2$).

5 Calculate the number of moles present in 3.2 kg of calcium carbonate ($CaCO_3$).

Ideal gas equation

Note: this topic is not assessed by CCEA.

The ideal gas equation is:

$$pV = nRT$$

where:
p is pressure measured in pascals (Pa)
V is volume measured in cubic metres (m^3)
n is the number of moles of gas
R is the gas constant ($R = 8.31 \, J \, K^{-1} \, mol^{-1}$)
T is temperature measured in kelvin (K).

When using this equation you must always look at the units of the data given. You may have to convert some of these.

 **A** Worked example

A sample of ammonia gas occupied a volume of 0.0143 m³ at 35 °C and a pressure of 100 kPa. The gas constant $R = 8.31\,\mathrm{J\,K^{-1}\,mol^{-1}}$. Calculate the amount, in moles, of ammonia in this sample.

- This equation involves use of the gas equation:

  $pV = nRT$

- If you need help in changing the subject of the equation see page 50.

Step 1: list all of the quantities given and convert them to the correct units.

    $p = 100\,\mathrm{kPa}$. The pressure must be in pascals. To convert from kPa to Pa you must multiply by 1000. $p = 100\,\mathrm{kPa} \times 1000 = 100\,000\,\mathrm{Pa}$

    $V = 0.0143\,\mathrm{m^3}$ — the units are correct

    $R = 8.31\,\mathrm{J\,K^{-1}\,mol^{-1}}$

    $T = 35\,°\mathrm{C}$. The temperature must be in kelvin. To convert from degrees Celsius to kelvin, add 273. Therefore, $T = 35\,°\mathrm{C} + 273 = 308\,\mathrm{K}$

Step 2: change the subject of the equation in order to find n.

    $pV = nRT$

    $nRT = pV$

Divide both sides by RT to get n on its own on the left of the equation.

$$\frac{n\cancel{RT}}{\cancel{RT}} = \frac{pV}{RT}$$

This simplifies to $n = \dfrac{pV}{RT}$

Step 3: substitute all the values, in the correct units.

$$n = \frac{100\,000 \times 0.0143}{8.31 \times 308} = 0.559\,\mathrm{mol}$$

B Guided question

Copy out the workings and complete the answers on a separate piece of paper.

1 **Calculate the volume of chlorine in dm³ present in 0.324 moles of chlorine at a temperature of 25 °C and a pressure of 100 kPa.**

This equation involves the use of the gas equation:

    $pV = nRT$

Step 1: list all of the quantities given and convert them to the correct units.

    $p = 100\,\mathrm{kPa}$. The pressure must be in pascals. To convert from kPa to Pa you must multiply by 1000. $p = 100\,\mathrm{kPa} \times 1000 = \underline{\hspace{2cm}}$.

    $R = 8.31\,\mathrm{J\,K^{-1}\,mol^{-1}}$

    $T = 25\,°\mathrm{C}$. The temperature must be in kelvin. To convert from Celsius to kelvin, add 273.

    $T = 25\,°\mathrm{C} + 273 = \underline{\hspace{2cm}}$

    $n = 0.324\,\mathrm{mol}$

Step 2: change the subject of the equation in order to find the volume.

$$pV = nRT$$

Divide both sides by p to get V on its own on the left of the equation.

$$\frac{\cancel{p}V}{\cancel{p}} = \frac{nRT}{p}$$

This simplifies to:

$$V = \frac{nRT}{p}$$

Step 3: substitute all the values, in the correct units.

Step 4: the answer will be in m^3 so to convert it to dm^3, multiply by 1000.

Ⓒ Practice questions

2   A sample of 0.136 mol of oxygen gas has a volume of 1822 cm^3 at a pressure of 104 kPa. What is the temperature of the gas? The gas constant $R = 8.31\,J\,K^{-1}\,mol^{-1}$.

3   A sample of magnesium nitrate decomposed to produce 0.502 mol of gas. Calculate the volume, in dm^3, that this gas would occupy at 323 K and 100 kPa. The gas constant $R = 8.31\,J\,K^{-1}\,mol^{-1}$.

4   In an experiment, 0.658 mol of SO_2 was produced. This gas occupied a volume of 0.0320 m^3 at a pressure of 100 kPa. Calculate the temperature of the SO_2 and state the units. The gas constant $R = 8.31\,J\,K^{-1}\,mol^{-1}$.

5   A sample of hydrogen sulfide gas occupied a volume of $1.63 \times 10^{-2}\,m^3$ at 37 °C and a pressure of 100 kPa. The gas constant $R = 8.31\,J\,K^{-1}\,mol^{-1}$. Calculate the amount, in moles, of hydrogen sulfide in this sample.

TIP

The volume 1.63×10^{-2} in question 5 is written in standard form. For notes on standard form see page 25.

Entropy

Note: this topic is for A-level candidates only.

The units of entropy are usually given in $J\,K^{-1}\,mol^{-1}$ but when calculating ΔG in $kJ\,mol^{-1}$, conversions to $kJ\,K^{-1}\,mol^{-1}$ must be used.

ΔG is calculated using the expression:

$$\Delta G = \Delta H - T\Delta S$$

where:

ΔG = change in Gibbs free energy in $kJ\,mol^{-1}$
ΔH = enthalpy change in $kJ\,mol^{-1}$
ΔS = entropy change in $kJ\,K^{-1}\,mol^{-1}$
T = temperature in K.

A Worked example

Glucose ($C_6H_{12}O_6$) is produced during respiration by the reaction of carbon dioxide and water.

$$6CO_2 + 6H_2O \rightarrow C_6H_{12}O_6 + 6O_2$$

$\Delta H = +2880\,kJ\,mol^{-1}$, $\Delta S = -256\,J\,K^{-1}\,mol^{-1}$

Calculate ΔG in $kJ\,mol^{-1}$ at $25\,°C$.

Step 1: list all of the quantities given and convert them to the correct units.

$\Delta H = +2880\,kJ\,mol^{-1}$. The units are correct.

$\Delta S = -256\,J\,K^{-1}\,mol^{-1}$. This must be converted to $kJ\,K^{-1}\,mol^{-1}$ by dividing by 1000.

$\Delta S = \dfrac{-256}{1000} = -0.256\,kJ\,K^{-1}\,mol^{-1}$

$T = 25\,°C$. This must be converted to K by adding 273.

Therefore $T = 25\,°C + 273 = 298\,K$.

Step 2: substitute the values into the equation.

$\Delta G = \Delta H - T\Delta S$

$\quad = +2880 - (298 \times -0.256)$

$\quad = +2880 - (-76.288)$

$\quad = +2880 + 76.288 = 2956.288\,kJ\,mol^{-1}$

TIP

 $+$ **Subtracting a negative number** is the same as **adding** so in the worked example above $2880 - (-76.288) = 2880 + 76.288$

Figure 1.5 Subtracting a negative number

B Guided question

Copy out the workings and complete the answers on a separate piece of paper.

1 **Calculate the temperature at which the value of $\Delta G = 0\,kJ\,mol^{-1}$ for the formation of ammonia in the Haber Process. $\Delta H = -46.2\,kJ\,mol^{-1}$, $\Delta S = -99.5\,J\,K^{-1}\,mol^{-1}$**

Step 1: list all of the quantities given and convert them to the correct units.

$\Delta H = -46.2\,kJ\,mol^{-1}$. The units are correct.

$\Delta S = -99.5\,J\,K^{-1}\,mol^{-1}$. This must be converted to $kJ\,K^{-1}\,mol^{-1}$ by dividing by 1000.

$\Delta S = \dfrac{-99.5\,J\,K^{-1}\,mol^{-1}}{1000} = \underline{\qquad}$

Step 2: substitute and find T by changing the subject of the equation.

$\Delta G = \Delta H - T\Delta S$

$0 = -46.2 - (T\Delta S)$

C Practice questions

2 For the reaction between NO_2 and O_3, $\Delta H = -198\,\text{kJ}\,\text{mol}^{-1}$ and $\Delta S = 180\,\text{J}\,\text{K}^{-1}\,\text{mol}^{-1}$. Calculate the value of ΔG at $30\,^\circ\text{C}$.

3 Sodium carbonate can decompose into sodium oxide and carbon dioxide

$$Na_2CO_3(s) \longrightarrow Na_2O(s) + CO_2(g)$$

For this reaction, $\Delta H = +323\,\text{kJ}\,\text{mol}^{-1}$ and $\Delta S = +153.7\,\text{J}\,\text{K}^{-1}\,\text{mol}^{-1}$. If $\Delta G = 0$, calculate the temperature in $^\circ\text{C}$.

Indices

For A-level chemistry, you must be able to work out units for rate constants and equilibrium constants. To help with this it is useful to understand the term index and the laws of indices.

The **index** (sometimes called the exponent or power) of a number states **how many times** to use the number in a multiplication. It is written as a superscript above the base number e.g. 2^3.

In 2^3, the index '3' means to use the number three times in a multiplication.

$2^3 = 2 \times 2 \times 2 = 8$

Or $5^4 = 5 \times 5 \times 5 \times 5 = 625$

Two special indices to remember are detailed below.

■ Any non-zero number raised to the **power of 1** is itself, for example, $3^1 = 3$ and $15^1 = 15$. The expression is $x^1 = x$.

■ Any non-zero number raised to the **power of 0** is 1, for example, $2^0 = 1$ and $5^0 = 1$. Remember $x^0 = 1$.

Multiplication of indices

When *multiplying* numbers which have indices, *add* the powers. Note that to do this the base number (in this case x) must be the same.

For example, $x^2 \times x^3 = x^{(2+3)} = x^5$

and $x^3 \times x^{-1} = x^{(3-1)} = x^2$

In terms of chemistry and working out units of equilibrium constants (K_c) and rate constants (k), you must be able to multiply concentration units ($\text{mol}\,\text{dm}^{-3}$).

Brackets

When taking the power of a number already raised to a power, multiply the powers, if the base number is the same.

$(x^2)^3 = x^{(2\times3)} = x^6$

Division

When dividing numbers with indices, subtract the indices:

$$\frac{x^5}{x^4} = x^{(5-4)} = x^1 = x$$

Remember when dividing it is possible to cancel, if the same term is on the top and bottom of the fraction.

$$\frac{\cancel{x^2} \times x^3}{\cancel{x^2}} = x^3$$

Reciprocal

The reciprocal of a number is $\dfrac{1}{\text{number}}$.

$$x^{-1} = \frac{1}{x}$$

 Worked examples

a **What are the units of $mol\,dm^{-3} \times mol\,dm^{-3}$?**

- Remember mol is really mol^1.
- It is often easier to collect similar terms together first, and then add the indices.

  $$mol \times mol \times dm^{-3} \times dm^{-3} = mol^{(1+1)}\,dm^{(-3\,+-3)} = mol^2\,dm^{-6}$$

- The units $mol\,dm^{-3}$ are concentration units, so another, simpler method is to replace the unit $mol\,dm^{-3}$ by the word concentration.

Step 1:

$$mol\,dm^{-3} \times mol\,dm^{-3} = \text{concentration} \times \text{concentration}$$

$$= (\text{concentration})^{(1+1)} = (\text{concentration})^2$$

Step 2: convert this back to units.

$$(\text{concentration})^2 = (mol\,dm^{-3})^2 = mol^2\,dm^{-6}$$

b **What are the units $\dfrac{(mol\,dm^{-3})^2}{mol\,dm^{-3}}$?**

$$\frac{(mol\,dm^{-3})^2}{(mol\,dm^{-3})} = \frac{(mol\,dm^{-3})^{\cancel{2}}}{\cancel{(mol\,dm^{-3})}} = mol\,dm^{-3}$$

$mol\,dm^{-3}$ is on the top and bottom of the fraction so they cancel out and this equals: $mol\,dm^{-3}$.

Or alternatively this can be written:

$$\frac{(\text{concentration})^2}{\text{concentration}} = \text{concentration}^{(2-1)} = \text{concentration} = mol\,dm^{-3}$$

c **What are the units of $\dfrac{(mol\,dm^{-3})^2}{(mol\,dm^{-3})^2}$?**

$$\frac{(mol\,dm^{-3})^2}{(mol\,dm^{-3})^2} = \frac{\cancel{(mol\,dm^{-3})^2}}{\cancel{(mol\,dm^{-3})^2}} = \text{no units}$$

as the units top and bottom are the same, they cancel each other out.

Alternatively,

$$\frac{(\text{concentration})^2}{(\text{concentration})^2} = \text{concentration}^{(2-2)} = \text{concentration}^0 = \text{no units}$$

d   What are the units of $\dfrac{(\text{mol dm}^{-3})^2}{(\text{mol dm}^{-3})^4}$ **?**

$$\frac{(\text{mol dm}^{-3})^2}{(\text{mol dm}^{-3})^4} = \frac{(\text{mol dm}^{-3})^2}{(\text{mol dm}^{-3})^{\cancel{4}\,2}} = \frac{1}{(\text{mol dm}^{-3})^2} = (\text{mol dm}^{-3})^{-2} = \text{mol}^{-2}\,\text{dm}^6$$

$$\frac{(\text{concentration})^2}{(\text{concentration})^4} = \frac{(\text{concentration})^2}{(\text{concentration})^{\cancel{4}\,2}} = \frac{1}{(\text{concentration})^2} = \text{concentration}^{-2} = (\text{mol dm}^{-3})^{-2}$$

$$= \text{mol}^{-2}\,\text{dm}^6$$

e   What are the units of $\dfrac{\text{mol dm}^{-3}\,\text{s}^{-1}}{(\text{mol dm}^{-3})^3}$ **?**

$$\frac{\text{mol dm}^{-3}\,\text{s}^{-1}}{(\text{mol dm}^{-3})^3} = \frac{\text{mol dm}^{-3}\,\text{s}^{-1}}{(\text{mol dm}^{-3})^{\cancel{3}\,2}} = \frac{\text{s}^{-1}}{(\text{mol dm}^{-3})^2} = \frac{\text{s}^{-1}}{\text{mol}^{(1\times2)}\,\text{dm}^{(-3\times2)}} = \frac{\text{s}^{-1}}{\text{mol}^2\,\text{dm}^{-6}} = \text{mol}^{-2}\,\text{dm}^6\,\text{s}^{-1}$$

f   The K_c **expression for the reaction of ethanol and ethanoic acid to produce ethyl ethanoate ($CH_3COOCH_2CH_3$) is:**

$$K_c = \frac{[CH_3COOCH_2CH_3][H_2O]}{[CH_3COOH][CH_3CH_2OH]}$$

What are the units of K_c**?**

The square brackets mean concentration so simply substitute concentration units into the expression.

$$\text{units of } K_c = \frac{(\text{mol dm}^{-3})^2}{(\text{mol dm}^{-3})^2} = \text{no units}$$

This is because the units are the same on the top and bottom and so they cancel each other out.

B Guided questions

Copy out the workings and complete the answers on a separate piece of paper.

Note: question 1 is for A-level candidates only.

1   For the general rate equation:

$$\text{rate} = k[\text{A}]^2[\text{B}]$$

find the units of k**.**

- The units of rate are $\text{mol dm}^{-3}\,\text{s}^{-1}$.

- The units of concentration [ ] are mol dm^{-3}.

Step 1: substitute the units into the equation and cancel.

$$\cancel{\text{mol dm}^{-3}}\,\text{s}^{-1} = k\,(\text{mol dm}^{-3})^2 \times (\cancel{\text{mol dm}^{-3}})$$

Step 2: change the subject of the equation by dividing both sides by $(\text{mol dm}^{-3})^2$. This makes k the subject.

2 **Write units for the equilibrium constant K_c, if $K_c = \dfrac{[CO_2][H_2]^4}{[CH_4][H_2O]^2}$.**

Step 1: the square brackets mean concentration so substitute concentration units into the expression.

$$K_c = \frac{[CO_2][H_2]^4}{[CH_4][H_2O]^2} = \frac{(mol\ dm^{-3})(mol\ dm^{-3})^4}{(mol\ dm^{-3})(mol\ dm^{-3})^2}$$

Step 2: cancel the $mol\ dm^{-3}$ on the top and bottom of the fraction.

$$K_c = \frac{\cancel{(mol\ dm^{-3})}(mol\ dm^{-3})^4}{\cancel{(mol\ dm^{-3})}(mol\ dm^{-3})^2} = \frac{(mol\ dm^{-3})^4}{(mol\ dm^{-3})^2}$$

Step 3: you can now cancel $(mol\ dm^{-3})^2$ on the top and bottom of the fraction.

Step 4: multiply out the brackets.

3 **The reaction $SO_2 + O_2 \longrightarrow SO_3$ has the following K_c expression:**

$$K_c = \frac{[SO_3]}{[SO_2][O_2]^{\frac{1}{2}}}$$

What are the units for K_c?

Step 1: the square brackets mean concentration so substitute concentration units into the expression.

$$K_c = \frac{(mol\,dm^{-3})}{(mol\,dm^{-3})(mol\,dm^{-3})^{\frac{1}{2}}}$$

Step 2: cancel the $mol\,dm^{-3}$ on the top and bottom.

This leaves:

$$K_c = \frac{1}{(mol\ dm^{-3})^{\frac{1}{2}}}$$

Step 3: multiply out the brackets on the bottom line (denominator).

Step 4: use the rule $\dfrac{1}{x} = x^{-1}$ to write the units.

C Practice questions

4 Find the units by simplifying the expressions below:

a $mol\,dm^{-3} \times mol\,dm^{-3}$

b $mol\,dm^{-3} \times (mol\,dm^{-3})^2$

c $(mol\,dm^{-3})^2 \times (mol\,dm^{-3})^2$

d $\dfrac{(mol\,dm^{-3})^2}{(mol\,dm^{-3})^3}$

e $\dfrac{(mol\,dm^{-3})^4}{(mol\,dm^{-3})^2}$

f $\dfrac{(mol\,dm^{-3})^2}{(mol\,dm^{-3}) \times (mol\,dm^{-3})^2}$

g $\dfrac{1}{(mol\,dm^{-3})^2}$

5   $K_c = \dfrac{[CO][H_2]^3}{[CH_4][H_2O]}$

Find the units of K_c.

Note: question 6 is for A-level candidates only.

6   For the following general rate equation: rate = $k[X]^x[Y]^y$

a   What are the units of rate?

b   What are the units of the rate constant if $x = 0$ and $y = 1$?

c   What are the units of the rate constant if $x = 1$ and $y = 2$?

7   For the equilibrium reaction $H_2 + I_2 \longrightarrow 2HI$, the K_c expression is:

$K_c = \dfrac{[HI]^2}{[H_2][I_2]}$

What are the units of K_c?

8   For the equilibrium reaction $\frac{1}{2}N_2 + \frac{3}{2}H_2 \longrightarrow 2NH_3$, the K_c expression is:

$K_c = \dfrac{[NH_3]^2}{[N_2]^{\frac{1}{2}}[H_2]^{\frac{3}{2}}}$

What are the units of K_c?

Expressions in decimal and ordinary form

Decimal places

When adding or subtracting data, decimal places are used to indicate the precision of the answer. The term 'decimal place' refers to the numbers after the decimal point.

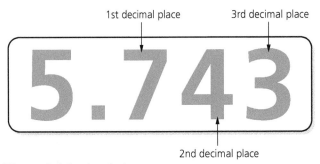

Figure 1.6 Decimal places

Sometimes in calculations you are asked to present your answer to one or two decimal places. To do this you need to round the number. For example:

- Rounding a number to one decimal place means there is only one digit after the decimal point.
- Rounding a number to two decimal places means there are two digits after the decimal point.

The rules for rounding are:

- If the next number is 5 **or more**, round up.
- If the next number is 4 **or less**, do not round up.

If you are rounding a number to two decimal places, for example, it is useful to underline all numbers up to two numbers after the decimal point. This then focuses your attention on the next number, which helps with rounding.

(A) Worked examples

a **Round 2.262 g to two decimal places.**

Step 1: underline all the numbers up to two numbers after the decimal point.

2.26̲2

Step 2: look at the number after the last underlined number. This number is 2 so you should follow the rule 'if the next number is 4 or less, do not round up'. This means the number 6 is unchanged.

The answer is 2.26 g (to 2 d.p.).

b **Round 4.9762 g to one decimal place.**

Step 1: underline all the numbers up to one number after the decimal point.

4.9̲762

Step 2: look at the number after the last underlined number. This number is 7 so you should follow the rule 'if the next number is 5 or more, round up'. This means the number 9 is rounded up to 10.

The answer is 5.0 g (to 1 d.p.).

(B) Guided questions

Copy out the workings and complete the answers on a separate piece of paper.

1 **Round 3.418 g to two decimal places.**

Step 1: underline all the numbers up to two numbers after the decimal point.

3.41̲8

Step 2: look at the number after the last underlined number and decide, using rounding rules, if you need to round up or not.

The answer is _____ (to 2 d.p.).

2 **In an experiment to find the mass of water removed on heating a solid, a student recorded the following measurements:**

Mass of hydrated solid + evaporating basin = 28.465 g

Mass of evaporating basin = 26.250 g

Mass of anhydrous solid + evaporating basin = 27.799 g

a **Calculate the mass of the anhydrous solid to two decimal places.**

Step 1: subtract the mass of the evaporating basin from the mass of the evaporating basin + anhydrous solid.

27.799 − 26.250 = 1.549 g

Step 2: underline the numbers up to two numbers after the decimal point.

1.54̲9

Step 3: look at the number after the last underlined number and decide, using rounding rules, if you need to round up or not.

b Calculate the mass of water removed to two decimal places.

Step 1: subtract the combined mass of the anhydrous solid + evaporating basin from the combined mass of the hydrated solid + evaporating basin.

28.465 − 27.799 =_____

Step 2: underline the numbers up to two numbers after the decimal point.

Step 3: look at the number after the last underlined number and decide, using rounding rules, if you need to round up or not.

Note: question 3 is for A-level candidates only.

3 **Calculate the pH of a $0.2\,\text{mol}\,\text{dm}^{-3}$ solution of hydrochloric acid. Record your answer to two decimal places.**

- To calculate the pH of a solution you need to use the following equation:

  $pH = -\log[H^+]$ where $[H^+]$ is the acid concentration.

  In this case $pH = -\log 0.2$.

- Remember that pH is $-\log[H^+]$ and the minus symbol means multiply by −1. Hence the answer for $\log[H^+]$ must be multiplied by −1. If $\log[H^+]$ gives a minus number this must be multiplied by −1 to give a positive pH value.

- If you are unsure how to use the log button on your calculator there are lots of examples on pages 59–61.

Step 1: find the log of 0.2 and write down all the numbers from your calculator. Then multiply by −1.

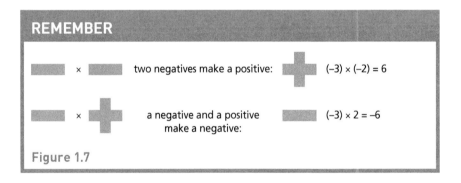

REMEMBER

| ▬ × ▬ | two negatives make a positive: | ╋ | $(-3) \times (-2) = 6$ |
| ▬ × ╋ | a negative and a positive make a negative: | ▬ | $(-3) \times 2 = -6$ |

Figure 1.7

Step 2: underline the numbers up to two numbers after the decimal point.

Step 3: look at the number after the last underlined number and decide, using rounding rules, if you need to round up or not.

ⓒ Practice questions

4 In an experiment different solids were heated in evaporating basins. To determine the mass of solid after heating, the mass of the evaporating basin must be subtracted from the combined mass of solid and evaporating basin. Calculate the mass of solid in grams to one decimal place by carrying out the following subtractions.

 a 30.25 − 28.53

 b 35.67 − 25.98

 c 24.34 − 22.23

5   Copy and complete Table 1.2.

Table 1.2

Mass/g	Mass recorded to two decimal places/g
29.883	
0.046	
32.6789	
13.999	
0.0894	
19 992.456	

Note: question 6 is for A-level candidates only.

6   Complete the following pH calculations and record the pH to one decimal place.

a   pH = −log 0.3

b   pH = −log 0.05

c   pH = −log 2.0

d   pH = −log 0.005

e   pH = −log 0.02

Accuracy

When adding or subtracting measurements with different numbers of decimal places, **the accuracy of the final answer can be no greater than the least accurate measurement.** This means that when measurements are added or subtracted, the answer should have the same number of decimal places as the smallest number of decimal places in any number involved in the calculation.

 Worked example

A student recorded the following masses of potassium chloride: 32.23 g, 2.1 g and 4.456 g. Calculate the total mass of potassium chloride. Give your answer to the appropriate precision.

Step 1: add the three masses.

32.23 + 2.1 + 4.456 = 38.786 g

Step 2: look at each of the masses and record the number of decimal places in each:

Table 1.3

Measurement	Number of decimal places
32.23 g	2
2.1 g	1
4.456 g	3

The number with least decimal places is 2.1. It has one decimal place, so the answer must be rounded to one decimal place.

Step 3: underline the numbers up to one number after the decimal point.

38.786

The number after the last underlined number is 8. This is greater than 5 so the answer must be rounded up.

The answer is 38.8 g.

B Guided question

Copy out the workings and complete the answers on a separate piece of paper.

1   **A student recorded some different temperatures in an experiment using thermometers with different accuracies. The results are shown in Table 1.4. Calculate the average temperature and give your answer to the appropriate number of decimal places.**

Step 1: record the number of decimal places in each reading in a copy of Table 1.4.

Table 1.4

Temperature/°C	Number of decimal places
10.2	
10	
10.3	
11	

Step 2: identify the number of decimal places in the temperature with least accuracy. This is the number of decimal places which should be in your answer.

Step 3: to calculate the average, add the four temperature readings and divide by the number of temperature readings (4). Record your answer to the correct number of decimal places.

C Practice questions

2   In an experiment different solids were heated in different evaporating basins. To determine the mass of solid after heating, the mass of the evaporating basin must be subtracted from the mass of solid and evaporating basin. Calculate the mass of solid used in each experiment to the appropriate number of decimal places, and copy and complete the table below.

Table 1.5

Mass of evaporating basin/g	Mass of evaporating basin and solid/g	Mass of solid (to appropriate number of decimal places)/g
34.567	23.4	
29.934	25.66	
25.49	22.1	
18.456	11.9	

3   A technician recorded the following masses of different chemicals. Work out the total mass of each chemical, giving your answer to the appropriate accuracy.
   a   Calcium carbonate 43.2 g, 0.245 g, 10.222 g
   b   Copper(II) sulfate 0.245 g, 10.393 g, 2.49 g
   c   Copper(II) oxide 3.23 g, 0.3439 g, 3.97 g

Standard form

Standard form is used to express very large or very small numbers so that they are more easily understood and managed. It is easier to say that a speck of dust weighs 1.2×10^{-6} g than to say it weighs 0.0000012 g or that a carbon to carbon bond has length 1.3×10^{-10} m than to say it is 0.00000000013 m.

Standard form must always look like this:

'A' must always be between 1 and 10

'n' is the number of places the decimal point moves

$A \times 10^n$

Figure 1.8 Standard form

'n' can also be thought of as the power of ten that A needs to be multiplied by to equal the original number.

Often when you carry out calculations using a calculator, the answer is displayed in standard form. A calculator will display standard form slightly differently, for example:

- 6.345×10^2 may be displayed as 6.345^2
- 4.573×10^{-3} may be displayed as 4.573^{-3}.

It is important that you can convert standard form back to ordinary form and ordinary form to standard form as shown in the worked examples below.

When converting from standard form to ordinary form, it is important that the number of significant figures is retained. (For notes on significant figures see page 36.)

 Worked examples

a   **Write 3 400 000 in standard form.**

Step 1: write the non-zero digits with a decimal point after the first number and then write '× 10^n' after it.

3.4×10^n

Step 2: count how many places the decimal point has moved to the left and write this value as the n value. Alternatively work out the number of times you need to multiply 3.4 by ten to get back to the original number.

$3\,400\,000 = 3.4 \times 10^6$

b   **Write 0.000932 in standard form.**

Step 1: write the non-zero digits with a decimal point after the first number and then write '× 10^n' after it.

9.32×10^n

Step 2: count how many places the decimal point has moved to the right (or alternatively the number of times you need to divide by 10 to get back to the original number) and write this value as the n value — the n is negative because the decimal point has moved to the right instead of the left.

$0.000932 = 9.32 \times 10^{-4}$

TIP

When entering standard form on your calculator, enter the decimal number and press the 'EXP' button on your calculator followed by the power. For example, to enter 7.3×10^{-9} on your calculator, enter 7.3 then enter 'EXP' then enter -9. If you are using the 'EXP' button it is incorrect to include the ×10.

Two numbers that you will use in chemistry which are usually presented in standard form are:

- Avogadro's number, 6.02×10^{23}
- the ionic product of water, K_w, 1×10^{14}. (Note: K_w is assessed at A-level only.)

B Guided questions

Copy out the workings and complete the answers on a separate piece of paper.

1 **Write 0.00645 in standard form.**

Step 1: write the non-zero digits with a decimal place after the first number and then write '$\times 10^n$' after it.

6.45×10^n

Step 2: count how many places the decimal point has moved to the right (or the number of times you need to divide by ten to get back to the original number) and write this value as the n value. It will be negative.

2 **Calculate the number of atoms present in 2.3 g of sodium.**

Step 1: calculate the number of moles of sodium using:

$$\text{moles} = \frac{\text{mass in g}}{M_r} = \frac{2.3}{23.0} = \underline{\qquad}$$

Step 2: one mole of atoms contain 6.02×10^{23} atoms. To find the number of atoms present in 2.3 g of sodium, multiply the amount in moles by 6.02×10^{23}.

C Practice questions

3 Write the following numbers in standard form.
   a 11 345
   b 3 234 567
   c 0.01
   d 0.00345
   e 0.00087
   f 1110.343
   g 9876.0089

4 Write the following numbers in ordinary form.
   a 3.2×10^6
   b 8.456×10^2
   c 5.6765×10^{-3}
   d 5.0×10^4
   e 4.655×10^{-6}

 Full worked solutions at www.hoddereducation.co.uk/essentialmathsanswers

    f 9.34×10^5

    g 238×10^{-6}

5 Use your calculator to calculate the following. Write your answer in ordinary form.

    a $0.032 \times 5 \times 10^4$

    b 0.00005×0.00003

    c $3.19 \times 10^5 \times 2 \times 10^{-2}$

    d $0.2 \times 10^{-4} \div 0.0098$

6 Use your calculator to calculate the following. Write your answer in standard form.

    a $6.02 \times 10^{23} \times 2.2 \times 10^{-12}$

    b $0.000034 \times 0.3333343$

    c $(3.2 \times 10^{34}) \div (2.1 \times 10^8)$

    d $0.002 \times 2.0 \times 10^3$

7 An atom of hydrogen contains a proton and an electron. Calculate the mass of a hydrogen atom if a proton has mass 1.6725×10^{-24} g and an electron has mass 9×10^{-28} g.

8 How many molecules are present in 8.505 g of $SiCl_4$?

9 How many atoms of sulfur are there in 0.0012 mol of S_8 molecules?

10 Calculate the number of tin atoms in 2.08 kg of tin.

Note: questions 11 and 12 are for A-level candidates only.

11 Calculate the pH of water at 25 °C when $K_w = 1.0 \times 10^{-14}$ mol^2 dm^{-6}. Use the equations pH $= -\log[H^+]$ and $K_w = [H^+][OH^-]$, remembering that in water $[H^+] = [OH^-]$ so $K_w = [H^+]^2$.

12 At 40 °C, $K_w = 2.92 \times 10^{-14}$ mol^2 dm^{-6}.

    Calculate the pH of water at 40 °C. Give your answer to two decimal places.

Ratios, fractions and percentages

Percentages

Per cent means 'out of 100'. If 50 per cent of the population own a car, this means 50 out of every 100 people have one. The symbol % means per cent. The concentration of some substances is sometimes written as '% by mass'.

For example, a label on a bottle of oven cleaner indicates that it contains 3.5% sodium hydroxide. This means that there are 3.5 g of sodium hydroxide in 100 g of solution. Percentages are a useful way of showing proportions. For example, the major uses of ammonia can be expressed as percentages as shown below.

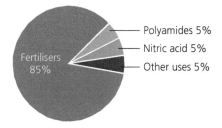

Polyamides 5%
Nitric acid 5%
Fertilisers 85%
Other uses 5%

Figure 1.9 The major uses of ammonia

A percentage is a fraction out of 100.

- 24% means 24 out of a hundred and is written as a fraction as $\frac{24}{100}$.
- 24% as a decimal is 0.24.

To convert a fraction or decimal to a percentage, simply multiply by 100.

(A) Worked examples

a **A 2.4 g sample of iron ore contains 1.8 g of Fe_2O_3. What percentage of the iron ore is Fe_2O_3?**

Step 1: express the quantity as a fraction.

$$\frac{1.8}{2.4}$$

Step 2: multiply by 100.

$$\frac{1.8}{2.4} \times 100 = 75\%$$

b **A sample of 3.0 g of limestone contains 30% of calcium carbonate. Calculate the mass of calcium carbonate present.**

Step 1: write the percentage as a fraction.

$$30\% \text{ is } \frac{30}{100}$$

Step 2: multiply the fraction by the quantity.

$$\frac{30}{100} \times 3.0 = 0.9 \, g$$

(B) Guided question

Copy out the workings and complete the answers on a separate piece of paper.

1 **Calculate the percentage of nitrogen in $Ca(NO_3)_2$.**

Step 1: find the M_r (relative molecular mass) of $Ca(NO_3)_2$.

$$M_r = 40.1 + (14.0 \times 2) + (16.0 \times 6) = \underline{\hspace{2cm}}$$

Step 2: there are two nitrogen atoms and so the mass of nitrogen in $Ca(NO_3)_2$ is:

$$14.0 \times 2 = \underline{\hspace{2cm}}$$

Step 3: express the quantity as a fraction.

$$\frac{\text{mass of nitrogen}}{M_r} = \underline{\hspace{2cm}}$$

Step 4: multiply by 100.

TIP
To find the percentage of a quantity: 1   Write the percentage as a fraction.
2   Multiply by the quantity.

 Practice questions

2   Calculate the percentage of:
  a   hydrogen in $Ca(OH)_2$
  b   potassium in $K_2Cr_2O_7$
  c   nitrogen in $(NH_4)_2SO_4$
  d   water in $CuSO_4.5H_2O$
  e   oxygen in $Na_2CO_3.10H_2O$.

3   Some egg shells were found to contain 35% of calcium carbonate. Calculate the mass of calcium carbonate in 2.3 g of egg shells. Give your answer to one decimal place.

4   A sample of 6.7 g of a rock was found to contain 4.1 g of silicon dioxide. What percentage of the rock was silicon dioxide?

Other types of calculation which involve percentages include percentage yield and atom economy calculations. To calculate these you need to recall and use the equations shown below.

$$\text{percentage yield} = \frac{\text{actual yield}}{\text{theoretical yield}} \times 100$$

$$\% \text{ atom economy} = \frac{\text{molecular mass of desired product}}{\text{sum of molecular masses of all reactants}} \times 100$$

Turn to page 97 to see some exam-style questions which use these equations.

Ratios

A ratio is a way to compare amounts of something. Recipes, for example, are sometimes given as ratios. To make pastry you may need to mix two parts flour to one part fat. This means the ratio of flour to fat is 2 : 1 and read as two to one. Ratios are written with a colon (:) between the numbers and usually only whole numbers are used.

Ratios are similar to fractions. They can both be simplified by finding common factors. Always try to divide by the highest common factor.

For example, in the ratio 12 : 15 the highest common factor (remember a factor is a number that divides into it exactly) is 3, so the ratio simplifies to 4 : 5.

Ratios are used in chemistry in many calculations, for example, in working out empirical formulae, in calculating reacting masses and in balancing equations.

Worked example

In an organic compound, the molecular formula is $C_2H_4O_2$. What is the empirical formula?

The empirical formula is the simplest whole number ratio of atoms present and is found by dividing the number of moles by the smallest of the numbers of atoms, in this case 2.

C	:	H	:	O	
$\frac{2}{2}$		$\frac{4}{2}$		$\frac{2}{2}$	
1		2		1	CH_2O is the empirical formula.

B Guided question

Copy out the workings and complete the answers on a separate piece of paper.

1 **A compound contains 0.050 moles of phosphorus and 0.125 moles of oxygen atoms. What is its molecular formula?**

Step 1: write down the elements present and the moles of each underneath.

P   : O

0.050 : 0.125

Step 2: to find the simplest ratio divide by the smaller of the number of moles.

$$\frac{0.050}{0.050} : \frac{0.125}{}$$

Sometimes you may not get a whole number ratio at this stage, and often multiplying by two or another number is necessary in order to get whole numbers.

C Practice questions

2 Calculate the empirical formula of:
  a $C_{16}H_{20}N_8O_4$
  b $Na_2S_2O_3$
  c $C_6H_{12}O_6$
  d P_4O_{10}

3 Calculate the empirical formula of a compound which contains:
  a 0.04 moles of $Al(NO_3)_3$ and 0.36 moles of water
  b 0.6 moles of lead atoms and 0.8 moles of oxygen atoms
  c 1.093 moles of chlorine atoms and 3.825 moles of oxygen atoms.

Balanced equations

In a balanced chemical equation the substances are all in ratio to each other, and this is shown by the numbers in front of each formula in the balanced symbol equation. For example, 2 moles of magnesium react with one mole of oxygen to produce 2 moles of magnesium oxide.

$$2Mg + O_2 \rightarrow 2MgO$$

The ratio is 2 moles Mg : 1 mole O_2 : 2 moles MgO.

Or in the reaction:

$$2Al + 6HCl \rightarrow 2AlCl_3 + 3H_2$$

The ratio between aluminium and hydrogen is $2Al : 3H_2$.

The ratio between aluminium and hydrochloric acid is $2Al : 6HCl$ which simplifies to $1Al : 3HCl$.

Ratios can be used in calculating the number of moles which react together in a reaction.

Worked examples

a **In the reaction**

$$2Pb(NO_3)_2 \text{ (s)} \rightarrow 2PbO \text{ (s)} + 4NO_2 \text{ (g)} + O_2 \text{ (g)}$$

how many moles of NO_2 are produced from 0.35 moles of $Pb(NO_3)_2$?

Step 1: write down the ratio between the two substances using the equation and simplify.

$Pb(NO_3)_2$	:	NO_2
2	:	4
1	:	2

Step 2: apply this ratio to the 0.35 moles of $Pb(NO_3)_2$.

There is twice as much NO_2 as $Pb(NO_3)_2$ so multiply $Pb(NO_3)_2$ moles by 2.

0.35	:	$(2 \times 0.35) = 0.70 \, mol$

b $4Al + 3O_2 \rightarrow 2Al_2O_3$

i **How many moles of aluminium are needed to produce 0.76 moles of aluminium oxide?**

Step 1: write down the ratio between the two substances using the equation and simplify.

Al	:	Al_2O_3
4	:	2
2	:	1

Step 2: apply this ratio to the 0.76 moles of aluminium oxide.

There are twice as many moles of aluminium so multiply by 2.

$0.76 \times 2 = 1.52 \, moles$ of Al

ii **How many moles of oxygen are needed to react completely with 0.2 moles of aluminium?**

Step 1: write down the ratio between the two substances.

Al	:	O_2
4	:	3

Step 2: divide by 4 to find how much oxygen reacts with 1 mole Al $= \frac{3}{4}$.

1	:	$\frac{3}{4}$
0.2	:	$0.2 \times \frac{3}{4} = 0.15 \, mol$

B Guided questions

Copy out the workings and complete the answers on a separate piece of paper.

1 **In the reaction**

$$N_2 + 3H_2 \rightarrow 2NH_3$$

how many moles of nitrogen are needed to react fully with 0.4 moles of hydrogen?

Step 1: write down the two species involved and the number of moles of each, from the equation ratio.

N_2	:	H_2
1	:	3

Step 2: there is three times as much H_2 as N_2, so divide H_2 moles (0.4) by 3.

2 **For the reaction**

$$P_4O_{10} + 6H_2O \rightarrow 4H_3PO_4$$

a **Calculate the amount, in moles, of water needed to react fully with 0.25 moles of P_4O_{10}.**

Step 1: write down the two species involved and the number of moles of each, from the equation ratio.

P_4O_{10}	:	H_2O
1	:	6

Step 2: there is six times as much H_2O as P_4O_{10}, so multiply the number of P_4O_{10} moles (0.25) by 6.

$$0.25 \times 6 = \underline{\hspace{2cm}}$$

b **Calculate the amount, in moles, of H_3PO_4 that will be produced when 0.3 moles of H_2O react.**

Step 1: write down the two species involved and the number of moles of each, from the equation ratio.

H_2O	:	H_3PO_4
6	:	4

Step 2: simplify this ratio first and then use it to find how much H_3PO_4 is formed from 0.3 moles.

C Practice questions

3 In the reaction

$$2Cu(NO_3)_2(s) \rightarrow 2CuO(s) + 4NO_2(g) + O_2(g)$$

a How many moles of O_2 are produced from 4 moles of $Cu(NO_3)_2$?

b How many moles of NO_2 are produced from 0.6 moles of $Cu(NO_3)_2$?

4 In the reaction

$$CaO + 3C \rightarrow CaC_2 + CO$$

a How many moles of carbon are needed to react completely with 0.33 moles of CaO?

b How many moles of CO are produced when 3.2 moles of carbon react?

5 In the reaction

$$3Pb + 2O_2 \rightarrow Pb_3O_4$$

a How many moles of oxygen are needed to react completely with 0.66 moles of Pb?

b How many moles of Pb_3O_4 are produced when 2.2 moles of oxygen react?

c How many moles of Pb_3O_4 are produced when 0.33 moles of lead react?

Estimating results

The effect of different conditions on K_c

Sometimes changing different experimental conditions has an effect on measurable values. Using maths, it is possible to estimate this effect.

For example, temperature is the only factor which affects the value of K_c for an equilibrium. It is not affected by changes in concentration or amount of substances present in the mixture, changes in pressure or the presence of a catalyst. Temperature affects the position of equilibrium and the concentrations of reactants and products will change as a result.

K_c is a fraction. A fraction is made of two parts: the top line which is called the numerator and the bottom line which is called the denominator. If the numerator is large and the denominator is small, then the fraction has a large value, for example, $\frac{10000}{0.2} = 50000$. However, if the numerator is small and the denominator is large then the fraction has a small value, for example, $\frac{0.2}{1000} = 0.0002$. In the K_c fraction, the concentration of products is in the numerator and the concentration of the reactants is in the denominator.

An increase in the concentration of the products (and decrease in the concentration of the reactants) will increase the value of K_c. This is because the numerator of the K_c expression gets larger and the denominator gets smaller, giving a larger value of K_c.

For a reaction: $aA + bB \rightleftharpoons cC + dD$

$$K_c = \frac{[C]^c[D]^d}{[A]^a[B]^b}$$

A decrease in the concentration of the products (and increase in the concentration of the reactants) will decrease the value of K_c when the temperature is changed.

Consider the equilibrium: $N_2(g) + 3H_2(g) \rightleftharpoons 2NH_3(g)$; $\Delta H = -92\,kJ\,mol^{-1}$

From your knowledge of Le Chatelier's principle, the forward reaction is exothermic (ΔH is negative) and so a *decrease* in temperature shifts the equilibrium position right in the forward exothermic direction to oppose the decrease in temperature, which means more ammonia is produced.

In the equation for K_c

$$K_c = \frac{[NH_3]^2}{[N_2][H_2]^3}$$

If more ammonia is produced the numerator (top) of the fraction gets bigger, and the denominator (bottom) gets smaller, as there is a smaller concentration of nitrogen and hydrogen, and so the value of K_c will increase.

> **TIP**
>
> Remember that the equilibrium constant is affected in this way by changes in temperature only. It does not change when the pressure or concentration of the system changes.

Worked example

A + 2B $\rightleftharpoons$ 2C; $\Delta H = -124$ kJ mol^{-1}

Estimate the effect, if any, of increasing the temperature on the value of K_c. The K_c expression is:

$$K_c = \frac{[C]^2}{[A][B]^2}$$

Step 1: decide if the reaction is exothermic or endothermic. ΔH is negative so the reaction is exothermic.

Step 2: decide the direction in which the position of equilibrium will move when the conditions change. If the temperature is increased in an exothermic reaction then the position of equilbrium will move to the left in the endothermic direction, to remove the heat.

Step 3: decide which part of the fraction is greater. There will be a greater concentration of reactants as the reaction moves left, and in the K_c expression the denominator increases and the numerator decreases, giving a smaller value of K_c.

B Guided questions

Copy out the workings and complete the answers on a separate piece of paper.

1 **PCl$_5$ $\rightleftharpoons$ PCl$_3$ + Cl$_2$; $\Delta H = +124$ kJ mol^{-1}**

   Explain the effect of increasing the temperature to 1000 K on the equilbrium constant K_c.

   Step 1: decide if the reaction is endothermic or exothermic.

   ΔH is positive so the reaction is endothermic.

   Step 2: use Le Chatelier's principle to decide what way the reaction will move to counteract the increase in temperature.

   Step 3: decide if there will be a greater concentration of products or reactants and note the effect on K_c by considering the size of the numerator (products) and denominator (reactants).

   $$K_c = \frac{[PCl_3][Cl]}{[PCl_5]}$$

2   **For the equilibrium reaction:**

$2X + Y \rightleftharpoons Z$; K_c is 25.2 at 150 °C and 4.3 at 800 °C.

$$K_c = \frac{[Z]}{[X]^2[Y]}$$

Explain whether the reaction is exothermic or endothermic.

Step 1: decide the effect of temperature on K_c. As the temperature increases, K_c decreases.

Step 2: if K_c is small, decide which is a bigger value, the numerator or the denominator.

    The _____ is bigger.

If there is a greater concentration of products, the equilibrium moves right on increasing temperature and is endothermic. If there is a greater concentration of reactants, the equilibrium moves left on increasing temperature and is exothermic.

Step 3: the equilibrium moves _____ on increasing the temperature and the reaction is _____.

Practice questions

3   Consider the reaction:

    $H_2 + I_2 \rightleftharpoons 2HI$; $\Delta H = +104 \, \text{kJ mol}^{-1}$

The value of K_c is 54.1 at a particular temperature.

Estimate the effect of increasing the temperature on the equilibrium position and on K_c.

4   For the equilibrium:

    $2SO_2(g) + O_2(g) \rightleftharpoons 2SO_3(g)$; $\Delta H = -197 \, \text{kJ mol}^{-1}$

Estimate the effect on K_c if the temperature is increased.

5   In a reaction, K_c decreases as temperature increases. Decide if the reaction is exothermic or endothermic.

6   For the Haber process, $N_2(g) + 3H_2(g) \rightleftharpoons 2NH_3(g)$; $\Delta H = -92 \, \text{kJ mol}^{-1}$, estimate the effect of increasing the temperature on the equilibrium position and on K_c.

2 Handling data

Significant figures

In calculations you may often get a long decimal answer on your calculator display. It is important to round this correctly using significant figures. Significant figures are those numbers which carry meaning and contribute to its precision.

The first significant figure of a number is the first digit that is not a zero.

The rules for significant figures are:

1 Always count non-zero digits. For example, 21 has two significant figures and 8.923 has four.

2 Never count zeros at the start of a number (leading zeros) even when there is a decimal point in the number. For example, 021, 0021 and 0.0021 all have two significant figures.

3 Always count zeros that fall between two non-zero digits. For example, 20.8 has three significant figures and 0.00103004 has six significant figures.

4 When a number with no decimal point ends in several zeros, these zeros may or may not be significant. The number of significant figures should then be stated. For example: 20 000 (to 3 s.f.) means that the number has been measured to the nearest 100 while 20 000 (to 4 s.f.) means that the number has been measured to the nearest 1000.

Figure 2.1 A very long decimal answer on a calculator can be rounded to significant figures. What is this number to three significant figures?

The rules are best illustrated with some examples.

■ 34.23 has four significant figures — always count non-zero digits.

■ 6000 has no decimal point and ends in several zeros so it is difficult to say if the zeros are significant. With zeros at the end of a number, the number of significant figures should be stated.

■ 2000.0 has a decimal place hence it has five significant figures.

■ 0.036 has two significant figures — never count the zeros at the start of a number even when there is a decimal point.

■ 3.0212 has five significant figures.

In calculations you should round the answer to a certain number of significant figures.

The rules for rounding are:
- if the next number is 5 or more, round up
- if the next number is 4 or less, do not round up.

 Worked example

Subtract 7.799 g − 6.250 g and give your answer to three significant figures.

Your calculation would yield 1.549 g. If the answer is given to three significant figures this would be 1.55 g, because 9 is greater than 5 so you need to round up.

B Guided question

Copy out the workings and complete the answers on a separate piece of paper.

1 **What is 3 478 906 to three significant figures?**

Step 1: underline the first three figures.

3478906

Step 2: write down the first two figures.

34

Step 3: look at the number after the one you have underlined and apply the rounding rules.

Step 4: put zero for all the remaining digits.

C Practice questions

2 What is the number of significant figures in each of the following?
   a 13.43   d 1.604
   b 2300   e 0.094
   c 2300.0

3 Write the following numbers to the stated number of significant figures.
   a 35 561.22 to 4 s.f.   d 442.45 to 4 s.f.
   b 5.278 to 3 s.f.   e 0.000045193 to 2 s.f.
   c 423 to 1 s.f.

Reporting calculations to an appropriate number of significant figures

When combining measurements with different degrees of accuracy and precision, **the accuracy of the final answer can be no greater than the least accurate measurement**. This means that when measurements are multiplied or divided, the answer can contain no more significant figures than the least accurate measurement. Often you will be asked to 'give your answer to the appropriate precision'.

Figure 2.2 A chain is as strong as its weakest link. A calculated answer is as accurate as the least accurate measurement in the calculation.

A Worked example

Calculate the value of $\dfrac{1.84 \times 2.3}{3.02}$. Give your answer to the appropriate precision.

Step 1: using a calculator the value is 1.401324.

Step 2: write down the number of significant figures in each measurement.

Table 2.1

Measurement	Number of significant figures
1.84	3
2.3	2
3.02	3

From the table you can see that the least accurate measurement is 2.3, which has two significant figures. Hence your answer should be rounded to two significant figures.

1.401324 rounds to 1.4 (to two significant figures).

B Guided question

Copy out the workings and complete the answers on a separate piece of paper.

1 In a titration, 20.5 cm³ of 0.25 mol dm⁻³ sodium hydroxide solution reacts with 1.2 mol dm⁻³ hydrochloric acid. Calculate the volume of hydrochloric acid required to neutralise the sodium hydroxide solution. Give your answer to the appropriate precision.

Step 1: write down the number of significant figures in each measurement.

Table 2.2

Measurement	Number of significant figures
20.5 cm³	
0.25 mol dm⁻³	
1.2 mol dm⁻³	

Step 2: write down how many significant figures there are in the least accurate measurement.

Step 3: calculate the number of moles of sodium hydroxide using $\dfrac{v \times c}{1000}$, and by ratio the number of moles of hydrochloric acid will be the same.

Step 4: use $v = \dfrac{n \times 1000}{c}$ to find the volume, where n is the number of moles of hydrochloric acid.

Step 5: round your answer to the same number of significant figures as the least accurate measurement.

C Practice questions

2 Use density = $\frac{mass}{volume}$ to calculate the density of a block of iron that has a mass of 40.52 g and volume of 5.1 cm³. Give your answer to an appropriate number of significant figures.

3 Calculate the number of moles in 2.2 g of calcium. Give your answer to an appropriate number of significant figures.

4 In a titration 26.5 cm³ of 0.200 mol dm⁻³ sodium hydroxide solution reacts with 0.300 mol dm⁻³ hydrochloric acid. Calculate the volume of hydrochloric acid required to neutralise the sodium hydroxide solution. Give your answer to the appropriate precision.

5 For the equilibrium reaction A + 2B ⇌ C, the equilibrium constant is 2.9 mol⁻² dm⁶ at 25 °C. If the equilibrium concentration of A is 0.2 mol dm⁻³ and the equilibrium concentration of C is 0.175 mol dm⁻³, calculate the equilibrium concentration of B. Give your answer to an appropriate number of significant figures.

6 When 25.00 g of 1.00 mol dm⁻³ sodium hydroxide was mixed with 25.00 g of 1.00 mol dm⁻³ hydrochloric acid, the temperature rose by 9.2 K. Assuming that the specific heat capacity of the solution is 4.18 J K⁻¹ g⁻¹ and the density of the solution is 1.00 g cm⁻³, calculate a value for the enthalpy change when one mole of water is formed, to an appropriate number of significant figures.

7 A mixture of 4.8 g of ethanoic acid and 0.120 mol of ethanol was allowed to reach equilibrium at 20 °C. A 25.0 cm³ sample of this mixture was titrated with sodium hydroxide. The ethanoic acid remaining in the sample reacted exactly with 4.00 cm³ of 0.400 mol dm⁻³ sodium hydroxide solution. When calculating K_c using all of these values, to how many significant figures should you give your answer? Explain your answer.

> In all these examples, even if you have not covered the chemistry involved in the calculation, you should be able to work out the appropriate number of significant figures to give in the answer.

Significant figures and standard form

For numbers in scientific form, to find the number of significant figures ignore the exponent (n number) and apply the usual rules.

For example, 6.2090×10^{28} has five significant figures and 1.3×10^2 has two significant figures.

The same number of significant figures must be kept when converting between ordinary and standard form.

For example:

- 0.0050 mol dm⁻³ = 5.0×10^{-3} mol dm⁻³ (2 s.f.)
- 40.06 g = 4.006×10^1 g (4 s.f.)

- $90.0\,g = 9.00 \times 10^1\,g$ (3 s.f.)
- $0.01070\,kg = 1.070 \times 10^{-2}$ (4 s.f.)

The number 260.99 rounded to:
- 4 s.f. is 261.0
- 3 s.f. is 261
- 2 s.f. is 260
- 1 s.f. is 300.

Using standard form makes it easier to identify significant figures.

In the example above, 261 has been rounded to the two significant figure value of 260. However, if seen in isolation, it would be impossible to know whether the final zero in 260 is significant (and the value to three significant figures) or insignificant (and the value to two significant figures).

Standard form, however, is unambiguous:
- 2.6×10^2 is to two significant figures
- 2.60×10^2 is to three significant figures.

 Worked example

Convert 0.002350 to standard form, ensuring that you retain the correct number of significant figures.

Step 1: note how many significant figures are present in 0.002350. Remember that the zeros after the point are not significant so there are four significant figures.

Step 2: to write the number in standard form, write the four significant figures, with a decimal place after the first number, and then write '$\times 10^n$' after it.

2.350×10^n

Step 3: count how many places the decimal point has moved to the right and write this value as the n value. The n is negative because the decimal point has moved to the right instead of the left.

$0.002350 = 2.350 \times 10^{-3}$

When to round off in multi-step calculations

- Rounding off should be left until the very end of the calculation.
- Rounding off after each step, and using this rounded figure as the starting figure for the next step, is likely to make a difference to the final answer. This introduces a rounding error. Rounding errors are often introduced in multi-step calculations.

This can be illustrated with the following example.

When 3.072 g of a metal carbonate, of M_r 100.1, is reacted with 50.0 cm³ of 1.0 mol dm⁻³ HCl(aq) (which is an excess), a temperature rise of 6.5 °C is obtained. The specific heat capacity of the solution is 4.18 J g⁻¹ K⁻¹. Calculate the value for the enthalpy change in kJ mol⁻¹.

$q = mc\Delta T$
$\Delta T = 6.5\,°C$
$q = mc\Delta T = 50.0 \times 4.18 \times 6.5 = 1358.5 = +1.3585\,kJ$

Since the least accurate measurement (the temperature rise) is only to two significant figures, the answer should also be quoted to two significant figures.

Therefore, the enthalpy change = 1.4 kJ

However, this figure is to be used subsequently to calculate the enthalpy change per mole so **the rounding off should not be applied until the final answer has been obtained**.

Number of moles of carbonate $= \dfrac{\text{mass}}{M_r} = \dfrac{3.072}{100.1} = 0.03068931$

The enthalpy change is +1.3585 kJ for 0.03068931 moles.

The enthalpy change per mole is $\dfrac{1.3585}{0.03068931} = +44.26622 = +44$ kJ to two significant figures.

Using the rounded value of 1.4 kJ for the heat produced gives an answer of

$\dfrac{1.4}{0.03068931} = 45.618 = +46$ kJ mol^{-1}. Hence there is a rounding error.

To avoid this do not round off until the final answer has been obtained.

B Guided questions

Copy out the workings and complete the answers on a separate piece of paper.

1 **Convert 2304.0 to standard form, ensuring that you retain the correct number of significant figures.**

Step 1: note how many significant figures are present in 2304.0. There are five.

Step 2: to write the number in standard form, write the five significant figures with a decimal place after the first number and then write '$\times 10^n$' after it.

2.3040×10^n

Step 3: count how many places the decimal point has moved to the left and write this value as the n value.

$2304.0 = \underline{\hspace{3cm}}$

Note that the last zero in 2304.0 is significant, hence when converting to standard form this must be retained.

2 **Complete the table to convert numbers from ordinary to standard form, retaining the number of significant figures.**

Step 1: decide on the number of significant figures present.

Step 2: write the number in standard form, retaining the number of significant figures.

Table 2.3

Number	Number of significant figures	Number in standard form
0.0060	2	6.0×10^{-3}
50.08		
30.0 g		
0.04070		

 Practice questions

3 Write the following in standard form, giving your answer to the same number of significant figures as presented below.

a 0.050

b 0.12

c 1 230 010.0

d 14 050.0

e 30.03

4 To determine the enthalpy change of combustion of ethanol, 100.0 cm³ of water was poured into a beaker and heated by combusting the ethanol. A temperature rise of 12.5 °C was recorded. The specific heat capacity of the solution is 4.18 J K⁻¹ g⁻¹ and the density of the solution is 1.00 g cm⁻³. Which one of the following uses the appropriate number of significant figures and correct standard form to represent the amount of energy transferred to the water?

A 5.225×10^3 J

B 5.23×10^3 J

C 52.3×10^2 J

D 52.0×10^2 J

Arithmetic mean

The arithmetic mean is found by adding together all the values and dividing by the total number of values. It may be referred to as the 'average' or simply as the mean.

 Worked example

The temperature of a solution was measured every 30 seconds for 3 minutes and the results recorded below.

Table 2.4

Time/s	0	30	60	90	120	150	180
Temperature/°C	21	22	23	24	24	23	22

Calculate the mean temperature.

In this case there are seven values, so the method is to add up the individual values and divide by 7.

$$\text{Mean} = \frac{21 + 22 + 23 + 24 + 24 + 23 + 22}{7} = 22.71 = 23 \text{ (to 2 s.f.)}$$

Calculating mean titres

During a titration, both initial and final burette readings are recorded, usually to two decimal places and a titre is calculated from the difference. A rough titration is carried out to determine the approximate amount of titre needed. To calculate the mean titre, **do not** include the rough titration value.

A **concordant titre** is obtained when the titres are within ±0.10 cm³ of each other.

Titres of 23.60 and 23.70 are concordant; titres of 23.60 and 23.85 are not concordant.

To calculate the mean of a set of results **only use concordant values**. A value which is not concordant, and distant from the other results, is called an **outlier**.

Titration tables can be presented in a range of different ways. Make sure you read the table carefully.

(A) Worked example

In a titration the results were recorded in Table 2.5. Calculate the mean titre.

Table 2.5

	Rough	Titration 1	Titration 2	Titration 3
Final burette reading/cm³	26.10	25.20	25.45	25.15
Initial burette reading/cm³	0.00	0.10	0.00	0.00
Titre/cm³	26.10	25.10	25.45	25.15

Titre 2 is not concordant and is not used to calculate the mean titre. The value 25.45 cm³ can be referred to as an 'outlier'.

$$\text{Mean} = \frac{25.10 + 25.15}{2} = 25.13 \text{ cm}^3$$

(B) Guided question

Copy out the workings and complete the answers on a separate piece of paper.

1 **In a titration the titres obtained were 20.35, 19.05, 19.10, 19.00. Calculate the mean titre.**

Step 1: examine the titres and determine if there are any outliers — results which are not concordant.

20.35 is an outlier, as it is distant from the other results, the other three are within ±0.10 cm³ of each other.

Step 2: add together the three concordant titres.

19.05 + 19.10 + 19.00 = _____

Step 3: divide this sum by the number of concordant titres (3).

(C) Practice questions

2 For each of the following titration tables (Table 2.6–2.8), calculate the mean titre and identify any outliers.

a   Table 2.6

	Rough	Titration 1	Titration 2	Titration 3
Initial burette reading/cm³	0.00	14.00	0.00	15.30
Final burette reading/cm³	13.00	26.50	12.45	28.00
Titre/cm³				

b   Table 2.7

	Rough	Titration 1	Titration 2	Titration 3
Final burette reading/cm³	4.70	8.65	11.85	16.80
Initial burette reading/cm³	0.20	4.65	7.65	12.85
Titre/cm³				

c   Table 2.8

	Rough	Titration 1	Titration 2	Titration 3	Titration 4
Final burette reading/cm³	23.16	45.40	22.55	45.20	22.50
Initial burette reading/cm³	0.01	23.15	0.25	22.50	0.30
Titre/cm³					
Concordant titres (✓)					

Calculating weighted mean

A weighted mean is where some values contribute more to the mean than others.

Relative atomic mass is a weighted mean of isotopic masses. The relative atomic mass of an element can be calculated from the relative isotopic masses of the isotopes (which are the same as the mass numbers) and the relative proportions in which they occur (abundance).

$$A_r = \frac{\sum(\text{mass of isotope} \times \text{relative abundance})}{\sum \text{relative abundance}}$$

where Σ represents the 'sum of' for all isotopes.

(A) Worked example

Calculate the relative atomic mass of rubidium to two decimal places.

	Relative isotopic mass	Abundance/%
Rb⁸⁵	85.00	72.15
Rb⁸⁷	87.00	27.85

- The relative atomic mass (A_r) is the weighted mean of the mass numbers so it takes into account the abundance of each isotope.
- To calculate the relative atomic mass you need to use the formula:

$$A_r = \frac{\sum(\text{mass of isotope} \times \text{relative abundance})}{\sum \text{relative abundance}}$$

Step 1: multiply the mass of each isotope (relative isotopic mass) by the abundance.

Rb^{85} mass of isotope × relative abundance

$= 85.00 \times 72.15 = 6132.75$

Rb^{87} mass of isotope × relative abundance

$= 87.00 \times 27.85 = 2422.95$

Step 2: insert these figures into the equation.

$$A_r = \frac{(85.00 \times 72.15) + (87.00 \times 27.85)}{72.15 + 27.85}$$

$$= \frac{6132.75 + 2422.95}{100.00}$$

$$= 85.557$$

$$= 85.56 \text{ to two decimal places}$$

B Guided question

Copy out the workings and complete the answers on a separate piece of paper.

1 **Calculate the relative atomic mass of magnesium to one decimal place.**

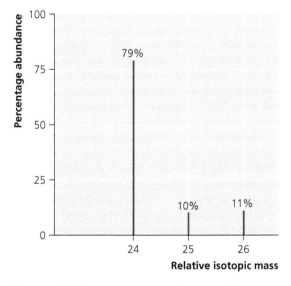

Figure 2.3 Mass spectrum of magnesium

The data is presented on a mass spectrum rather than in a table but the method is the same.

Step 1: find the sum of the abundances (total abundance) which is given by the height of each peak.

$79 + 10 + 11 = $ _____

Step 2: calculate relative atomic mass by multiplying each relative isotopic mass value by its abundance and dividing by the total abundance.

$$\frac{(24 \times 79) + (25 \times 10) + (26 \times 11)}{\text{total abundance}} = \underline{\qquad}$$

Practice questions

2 The table below shows the relative abundance of the four main isotopes of calcium. Calculate the relative atomic mass of calcium to two decimal places.

Table 2.9

Isotope	^{40}Ca	^{42}Ca	^{43}Ca	^{44}Ca
Percentage abundance/%	96.9	0.6	0.2	2.3

3 Iron has four isotopes as shown in the table. Calculate the relative atomic mass of iron.

Table 2.10

Isotope	^{54}Fe	^{56}Fe	^{57}Fe	^{58}Fe
Percentage abundance/%	5.85	91.76	2.12	0.28

4 The isotopes of sulfur and their abundance are shown in the table. Calculate the relative atomic mass of sulfur to two decimal places.

Table 2.11

Isotope	Percentage abundance/%
^{32}S	95.02
^{33}S	0.76
^{34}S	4.22

Identifying uncertainties in measurements

Uncertainty is an estimate attached to a measurement which gives the range of values within which the true value is thought to lie. This is normally expressed as a range of values such as 52.0 ±0.2.

For example, if the mass of a conical flask is measured with a balance that reads to 0.1 g, the reading, with uncertainty, is reported as 35.3 ±0.1 g. This means the 'actual' mass is known to lie between 35.2 and 35.4 g. If the same flask is measured on a balance that reads to 0.01 g, the mass might be reported as 35.28 ±0.01 g. So in this case, the 'actual' mass is known to be between 35.27 and 35.29 g. Thus, in the second mass reading, there is less uncertainty.

When ungraduated glassware, such as a volumetric flask or a pipette, is manufactured, the maximum uncertainty is usually marked on the glassware as shown in Figure 2.4.

Figure 2.4 A 15 ml pipette. The uncertainty is marked at ±0.03.

It is useful to calculate the percentage uncertainty in any result. This is often called the percentage error.

$$\text{percentage uncertainty (percentage error)} = \frac{\text{uncertainty} \times 100}{\text{quantity measured}}$$

 Worked example

A balance is used to weigh 4.34 g of magnesium and then another 23.34 g of magnesium. The uncertainty associated with the two decimal place balance reading is ±0.005 g. Calculate the percentage uncertainty for each result.

$$\text{percentage uncertainty} = \frac{0.005 \times 100}{4.34} = 0.12\%$$

$$\text{percentage uncertainty} = \frac{0.005 \times 100}{23.34} = 0.02\%$$

Notice that the percentage uncertainty is more significant when weighing out a smaller mass.

Multiple measurements

When several measurements are made, and the quantities measured by difference (subtraction), there will be an uncertainty in each measurement, which must be combined to give the uncertainty in the final value. This is the case in a titration and also in many heating to constant mass calculations. For multiple measurements using a balance with uncertainty ±0.005 g, there will be a maximum uncertainty of ±0.005 g for each measurement. For two mass measurements that give a resultant mass by difference, there are two uncertainties.

The formula for the percentage uncertainty is then:

$$\text{percentage uncertainty} = \frac{2 \times \text{uncertainty in each measurement} \times 100}{\text{quantity measured}}$$

 Worked example

In an experiment to find the mass of water lost on heating a solid, the following results were obtained. Calculate the mass of water lost and the percentage uncertainty in this value.

Mass of crucible + solid before heat = 25.45 g; uncertainty = 0.005 g

Mass of crucible + solid after heat = 24.21 g; uncertainty = 0.005 g

overall uncertainty = 2 × 0.005 g

mass lost = 1.24 g

percentage uncertainty in mass loss = $\dfrac{2 \times 0.005 \times 100}{1.24} = 0.81\%$

You must be also able to determine the uncertainty when two burette readings are used to calculate a titre value as shown in the Guided question on page 48.

B Guided question

Copy out the workings and complete the answers on a separate piece of paper.

1 **A burette has uncertainty $\pm 0.05\,\text{cm}^3$. In a titration, the initial burette reading was $0.05\,\text{cm}^3$ and the final burette reading was $22.55\,\text{cm}^3$. What is the percentage uncertainty in the titre value?**

Step 1: calculate the titre value using

titre value = final burette reading – initial burette reading = _____

Step 2: calculate the uncertainty. The overall uncertainty in any volume measured in a burette always comes from the two measurements, so to find the overall uncertainty multiply the uncertainty by two.

Overall uncertainty = _____

Step 3: calculate the percentage uncertainty using

$$\text{percentage uncertainty} = \frac{\text{uncertainty} \times 100}{\text{titre value}}$$

TIP

Be careful when asked to work out percentage uncertainty for multiple readings. Sometimes the question will give the **overall uncertainty** of an instrument and then state '**this uncertainty takes into account multiple measurements**'. If this is the case you do not need to work out the overall uncertainty.

C Practice questions

2 The uncertainty associated with a volumetric flask reading is $\pm 0.2\,\text{cm}^3$. What is the percentage uncertainty when making up a $250\,\text{cm}^3$ standard solution in this volumetric flask.

3 The uncertainty associated with a pipette reading is $\pm 0.06\,\text{cm}^3$. Calculate the percentage uncertainty when using a $25.0\,\text{cm}^3$ pipette.

4 What is the percentage error when a volume of $22.35\,\text{cm}^3$ is measured with a burette which has an error of $\pm 0.05\,\text{cm}^3$.

5 In an experiment a student made two different measurements. She measured $25\,\text{cm}^3$ of hydrochloric acid using a measuring cylinder with uncertainty $\pm 1\,\text{cm}^3$ and $0.40\,\text{g}$ of calcium carbonate using a balance with uncertainty $\pm 0.01\,\text{g}$. State and explain whether the balance or the measuring cylinder contributed most to the measurement errors in this experiment.

6 Which of the following pieces of apparatus has the lowest percentage error in the measurement shown?
   A Volume of $25.0\,\text{cm}^3$ measured with a burette with an error of $\pm 0.1\,\text{cm}^3$.
   B Volume of $20\,\text{cm}^3$ measured with a measuring cylinder with an error of $\pm 1\,\text{cm}^3$.
   C Mass of $0.320\,\text{g}$ measured with a balance with an error of $\pm 0.001\,\text{g}$.
   D Temperature change of $83.2\,^\circ\text{C}$ measured with a thermometer with an error of $\pm 0.1\,^\circ\text{C}$.

7   In an experiment to determine the temperature change in a reaction the initial temperature was 22.3 °C and the final temperature was 35.3 °C. The maximum total error of the thermometer is ± 0.1 °C. The error takes into account multiple measurements made from the thermometer. Calculate the percentage error in the temperature change.

8   A burette has uncertainty ± 0.05 cm^3. In a titration, the initial burette reading was 0.05 cm^3 and the final burette reading is 12.45 cm^3. Calculate the percentage uncertainty in the titre.

9   3.20 g of magnesium was weighed on a balance with error ± 0.001. It was heated and reweighed and a mass of 4.34 g was obtained. Calculate the percentage error in the change in mass.

10 In an experiment the mean titre is 19.45 cm^3. The error in the mean titre for this experiment is ± 0.15 cm^3. Calculate the percentage error in this mean titre.

3 Algebra

Understanding symbols

You need to be familiar with different symbols which can be used in mathematical equations or in chemical equations.

Table 3.1

Symbol	Meaning	Comment
=	is equal to	This is the algebraic symbol you will use most.
«	is much less than	
»	is much greater than	
<	is less than	
>	is greater than	
~	is similar to	
∝	is proportional to	
⇌	a reversible reaction	You will come across this symbol most often in equilibrium reactions, where it is used instead of an arrow to show that a reaction can occur in both directions. e.g. $N_2 + 3H_2 \rightleftharpoons 2NH_3$

Ⓐ Practice question

1   Decide if the following are true or false.
   a $10 < 5$
   b $15 > 9$
   c $10 = 10$
   d $10\,099 \gg 10$
   e $0.000023 \ll 23$
   f $3.2 \times 10^2 = 3200$
   g $5.45 \times 10^{-2} > 0.0004$
   h $989\,000 = 9.89 \times 10^3$

Changing the subject of an equation

An equation shows that two things are equal. It will have an equals sign '='.

This means that what is on the left of the equals sign is equal to what is on the right. An example of an equation is:

$$\frac{\text{mass (g)}}{M_r} = \text{moles}$$

The subject of an equation is the single variable (usually on the left of the '=') that everything else is equal to. In the example above the subject is 'moles'.

To help with chemical calculations it is very useful to be able to rearrange an equation and so change the subject of the equation.

A Worked examples

a **Make x the subject of the equation $y = x + 6$.**

Step 1: switch sides to get the new subject on the left.

$x + 6 = y$

Step 2: to get x by itself on the left-hand side you need to subtract 6 from the left side but, to keep the equation true, you need to subtract 6 from the right side as well.

$x + 6 - 6 = y - 6$

Step 3: simplify.

$x = y - 6$

b **Make mass the subject of the equation, $\text{moles} = \dfrac{\text{mass (g)}}{M_r}$.**

Step 1: switch sides to get the new subject on the left.

$\dfrac{\text{mass (g)}}{M_r} = \text{moles}$

Step 2: to get mass by itself on the left-hand side you need to remove M_r by multiplying both sides by M_r and cancelling the M_r on the left.

$\dfrac{\text{mass} \times \cancel{M_r}}{\cancel{M_r}} = \text{moles} \times M_r$

$\text{mass} = \text{moles} \times M_r$

B Guided questions

Copy out the workings and complete the answers on a separate piece of paper.

1 **Make x the subject of the equation $y - 3 = 3x + 6$.**

Step 1: switch sides to get the new subject on the left.

$3x + 6 = y - 3$

Step 2: to get $3x$ by itself on the left-hand side so you need to subtract 6 from the left side but, to keep the equation true, you need to subtract 6 from the right side as well, and then simplify.

$3x + 6 - 6 = y - 3 - 6$

$3x = \underline{\hspace{3cm}}$

Step 3: to get x by itself as the subject you need to divide the left side by three and, to keep the equation true, you must also divide the right side by three.

2 **Make $[CH_3COOC_2H_5]$ the subject of the equation**

$$K_c = \frac{[CH_3COOC_2H_5][H_2O]}{[C_2H_5OH][CH_3COOH]}.$$

Step 1: switch sides to get the new subject on the left.

$$\frac{[CH_3COOC_2H_5][H_2O]}{[C_2H_5OH][CH_3COOH]} = K_c$$

Step 2: to get $[CH_3COOC_2H_5]$ on the left you need to remove the term it is divided by $([C_2H_5OH][CH_3COOH])$ by multiplying both sides by it.

$$\frac{[CH_3COOC_2H_5][H_2O] \times [C_2H_5OH][CH_3COOH]}{[C_2H_5OH][CH_3COOH]} = K_c[C_2H_5OH][CH_3COOH]$$

Step 3: simplify and then to get $[CH_3COOC_2H_5]$ on its own you need to divide by $[H_2O]$.

Note: question 3 is for A-level candidates only.

3 **Rearrange the rate equation, rate = $k[NO]^2[O_2]$ to make [NO] the subject.**

Step 1: switch sides to get the new subject on the left.

$k[NO]^2[O_2]$ = rate

Step 2: to get [NO] by itself on the left-hand side you need to remove the $k[O_2]$. To do this you need to divide both sides by $k[O_2]$.

$$\frac{k[NO]^2[O_2]}{k[O_2]} = \frac{\text{rate}}{k[O_2]}$$

Step 3: simplify and you will get an expression for $[NO]^2$. You then need to take the square root of each side to get an expression for [NO].

TIP

The inverse operation is the operation that reverses the effect of another operation. For example, addition and subtraction are inverse operations. Squaring a number and taking the square root are also inverse operations. To rearrange an equation with a squared power use the inverse operation, which is a square root. For example, $3^2 = 9$ and $\sqrt{9} = 3$. In general $a^2 = b$ so $a = \sqrt{b}$.

C Practice questions

4 Rearrange the following equations to make x the subject.
   a $y = 2x + 1$
   b $3x = 4 + y$
   c $x + y = 8$
   d $y = mx + c$
   e $3 = 4x + 2y$
   f $2y + 3 = 2 - x$
   g $y(x + 1) = 3$
   h $y = xz^2$
   i $w = yx^2$

5 Rearrange the equations below to make the variable in **bold** the subject.

   a $\text{moles} = \dfrac{\textbf{mass}}{M_r}$

   b $\text{moles} = \dfrac{\textbf{vol (dm}^3\textbf{)}}{24}$

   c $PV = n\textbf{\textit{RT}}$

   d $\text{percentage yield} = \dfrac{\text{actual yield} \times 100}{\textbf{theoretical yield}}$

e $\quad v = \dfrac{n \times c}{1000}$

f $\quad K_c = \dfrac{[A][B]}{[D][C]^2}$

g $\quad K_c = \dfrac{[C]^2}{[A]^2[B]}$

h $\quad$ Rate $= k[A][B][C]$

i $\quad$ Rate $= k[A]^2$

j $\quad \Delta G^\circ = \Delta H^\circ - T\Delta S^\circ$

Solving algebraic equations

When solving algebraic equations, simply substitute the numerical values given in the question. Hess's law calculations often involve substituting numerical values into one of the following equations:

- $\Delta H_f = \sum \Delta H_c$ reactants $- \sum \Delta H_c$ products

- $\Delta H_c = \sum \Delta H_f$ products $- \sum \Delta H_f$ reactants

- $\Delta H_r = \sum \Delta H_f$ products $- \sum \Delta H_f$ reactants

where '$\sum$' means the 'sum of' and so all the values are added together.

A Worked example

Calculate a value for the standard enthalpy of formation of butane given the standard enthalpy changes of combustion shown in the table.

The equation for the standard enthalpy of formation of butane is:

$\quad 4C(s) + 5H_2(g) \rightarrow C_4H_{10}(g)$

Table 3.2

	ΔH_c/kJ mol^{-1}
C_4H_{10} (g)	−2876.5
C(s)	−393.4
H_2(g)	−285.8

- To calculate the enthalpy of formation, given combustion values, use the equation:

$\quad \Delta H_f = \sum \Delta H_c$ reactants $- \sum \Delta H_c$ products

Step 1: the reactants are $4C(s) + 5H_2(g)$.

$\quad \sum \Delta H_c$ reactants $= (4 \times -393.4) + (5 \times -285.8) = -1573.6 + (-1429.0) = -3002.6$ kJ mol^{-1}

Step 2: the product is $C_4H_{10}(g)$.

$\quad \sum \Delta H_c$ products $= -2876.5$ kJ mol^{-1}

Step 3:

$\quad \Delta H_f = \sum \Delta H_c$ reactants $- \sum \Delta H_c$ products

$\quad\quad = -3002.6 - (-2876.5)$

$\quad\quad = -3002.6 + 2876.5 = -126.1$ kJ mol^{-1}

- **Subtracting a negative number** is the same as **adding**, so in this example
$-3002.6 - (-2876.5) = -3002.6 + 2876.5 = -126.1$ kJ mol^{-1}

B Guided question

Copy out the workings and complete the answers on a separate piece of paper.

1   **Table 3.3 shows some standard enthalpies of formation.**

Table 3.3

	ΔH_f/kJ mol^{-1}
$NH_3(g)$	−46.0
$O_2(g)$	0.0
$N_2O(g)$	82.0
$H_2O(l)$	−286.0

Use these values to calculate the standard enthalpy change for the reaction:

$$2NH_3(g) + 2O_2(g) \rightarrow N_2O(g) + 3H_2O(l)$$

To calculate the enthalpy change of reaction use the equation:

$$\Delta H_r = \Sigma \Delta H_f \text{ products} - \Sigma \Delta H_f \text{ reactants}$$

Step 1: the reactants are $2NH_3(g)$ and $2O_2(g)$.

$$\Sigma \Delta H_f \text{ reactants} = (2 \times -46.0) + (2 \times 0.0) = -92 \text{ kJ mol}^{-1}$$

Step 2: the products are $N_2O(g)$ and $3H_2O(l)$.

$$\Sigma \Delta H_f \text{ products} = (1 \times 82.0) + (3 \times -286.0) = -776.0 \text{ kJ mol}^{-1}$$

Step 3: $\Delta H_r = \Sigma \Delta H_f \text{ products} - \Sigma \Delta H_f \text{ reactants}$

$$= -92 - (-776) = \underline{\hspace{2cm}} \text{ kJ mol}^{-1}$$

C Practice questions

2   The standard enthalpies of combustion of carbon (C), hydrogen (H_2) and propane (C_3H_8) are −394, −286 and −2219 kJ mol^{-1}, respectively. Use these values to calculate the standard enthalpy of formation of propane ($3C(s) + 4H_2(g) \rightarrow C_3H_8(g)$) in kJ mol^{-1}.

3   Use the standard enthalpies of formation given in the table below to calculate the standard enthalpy of combustion for butan-1-ol.

$$C_4H_9OH(l) + 6O_2(g) \rightarrow 4CO_2(g) + 5H_2O(g)$$

Table 3.4

	ΔH_f/kJ mol^{-1}
C_4H_9OH	−327
CO_2	−394
H_2O	−286
O_2	0

4   Calculate the standard enthalpy change of formation of hexane using the standard enthalpy changes of combustion in the table below. The equation is

$$6C(s) + 7H_2(g) \rightarrow C_6H_{14}(g)$$

Table 3.5

	ΔH_c/kJ mol^{-1}
C	−394
C_6H_{14}	−4163
H_2	−286

Sometimes, when solving algebraic equations you must first change the subject and then substitute the numerical values given in the question.

 Worked example

a **Calculate the amount, in moles, of calcium hydroxide present in 25.0 cm³ of a solution of concentration 0.25 mol dm⁻³.**

- The question gives a volume and a concentration and the number of moles is to be calculated.
- The equation to use is:
$$n = \frac{v \times c}{1000}$$
- If the volume of calcium hydroxide was given in dm³ then you would use the equation $n = v \times c$ and not divide by 1000.

The subject is amount in moles, so does not need to be changed. Simply substitute the numerical values and calculate the answer.

$$n = \frac{v \times c}{1000} = \frac{25.0 \times 0.25}{1000} = 0.0063 \text{ (to 2 s.f.)}$$

Note: example b is for A-level candidates only.

b **K_a is 1.74×10^{-5} mol dm⁻³ for ethanoic acid. Find the pH of a 0.20 mol dm⁻³ solution of ethanoic acid.**

- To find pH you need to use $\text{pH} = -\log[\text{H}^+]$.
- First you need to find $[\text{H}^+]$ so use:
$$K_a = \frac{[\text{H}^+]^2}{[\text{CH}_3\text{COOH}]}$$

Step 1: change the subject to $[\text{H}^+]$.

Switch sides to get the new subject on the left.

$$\frac{[\text{H}^+]^2}{[\text{CH}_3\text{COOH}]} = K_a$$

On the left-hand side you need the term $[\text{H}^+]$ on its own so multiply both sides by $[\text{CH}_3\text{COOH}]$

$$\frac{\cancel{[\text{CH}_3\text{COOH}]} \times [\text{H}^+]^2}{\cancel{[\text{CH}_3\text{COOH}]}} = K_a \times [\text{CH}_3\text{COOH}]$$

$$[\text{H}^+]^2 = K_a \times [\text{CH}_3\text{COOH}]$$

To remove the power of two on the left-hand side, you must take the square root of the right-hand side.

$$[\text{H}^+] = \sqrt{K_a \times [\text{CH}_3\text{COOH}]}$$

To find $[\text{H}^+]$ simply substitute the values into this equation. If K_a is 1.74×10^{-5} mol dm⁻³ and it is a 0.20 mol dm⁻³ solution of ethanoic acid then:

$$[\text{H}^+] = \sqrt{K_a \times [\text{CH}_3\text{COOH}]} = \sqrt{1.74 \times 10^{-5} \times 0.20} = \sqrt{3.48 \times 10^{-6}}$$
$$= 1.87 \times 10^{-3}$$

Step 2: substitute into the equation

$$\text{pH} = -\log[\text{H}^+] = -\log 1.87 \times 10^{-3}$$
$$\text{pH} = -(-2.7) = 2.7$$

B Guided questions

Copy out the workings and complete the answers on a separate piece of paper.

1 **Calculate the concentration of calcium hydroxide solution obtained when 0.0034 mole of calcium hydroxide is dissolved in 15.0 cm³ water.**

   - The information in the question is $n = 0.0034$ mol and $v = 15.0$ cm³. Hence use the equation:
     $$n = \frac{v \times c}{1000}$$

   - If the volume of calcium hydroxide was given in dm³ then you use the equation $n = v \times c$ and do not divide by 1000.

   - When substituting numerical values into an equation you must use the correct units for measurement.

   Step 1: the subject is the amount in moles, but you need to find the concentration, so the subject needs to be changed to c.

   Switch the sides so that c is on the left:
   $$\frac{v \times c}{1000} = n$$

   Step 2: to get c on its own on the left, remove the 1000 by multiplying both sides by 1000 and simplify.
   $$\frac{v \times c \times \cancel{1000}}{\cancel{1000}} = n \times 1000$$
   $$v \times c = n \times 1000$$

   Step 3: to remove v, divide both sides by v and simplify.
   Step 4: substitute the numerical values $n = 0.0034$ mol and $v = 15.0$ cm³ and calculate the answer.

2 **Rearrange the equilibrium expression to make [C] the subject.**
   $$K_c = \frac{[C]^2[D]}{[A][B]^2}$$

   Step 1: switch sides to get the new subject on the left.
   $$\frac{[C]^2[D]}{[A][B]^2} = K_c$$

   Step 2: to get [C] on its own on the left-hand side, multiply both sides by $[A][B]^2$ and cancel.

   $$\frac{[C]^2[D]}{\cancel{[A][B]^2}} \times \cancel{[A][B]^2} = K_c[A][B]^2$$

   $$[C]^2[D] = K_c[A][B]^2$$

   Step 3: to get $[C]^2$ on its own on the left you need to remove the [D] on the left, so divide both sides by [D].
   $$[C]^2 = \frac{K_c[A][B]^2}{[D]}$$

   Step 4: to remove the power of two in $[C]^2$ you must take the square root of both sides.

ⓒ Practice questions

3   K_c for a reaction is 44.0. Calculate the value of [HI], if [H_2] is 0.2 and [I_2] is 0.2.

$$K_c = \frac{[HI]^2}{[H_2][I_2]}$$

4   For an experiment, $K_c = \dfrac{[C]^2}{[A]^2[B]}$. Calculate the equilibrium value of [C] if

    $K_c = 44.9\,mol^{-1}\,dm^3$ and the equilibrium concentration of A is $1.17\,mol\,dm^{-3}$ and B is $0.34\,mol\,dm^{-3}$.

5   A 6.00 g sample of KCl was added to 25.0 g of water and the temperatures decreased from 21.0 °C to 13.4 °C. Calculate a value for the enthalpy of solution of KCl, assuming that the specific heat capacity of water is $4.18\,J\,K^{-1}\,g^{-1}$. Use $q = mc\Delta T$.

Note: question 6 is for A-level candidates only.

6   A rate equation is rate = $k[Y][Z]^2$. Calculate a value of the rate constant at this temperature if [Y] = $1.7 \times 10^{-2}\,mol\,dm^{-3}$, [Z] = $2.4 \times 10^{-2}\,mol\,dm^{-3}$, rate = $7.40 \times 10^{-5}\,mol\,dm^{-3}\,s^{-1}$. State the units.

The Arrhenius equation

Note: this topic is for A-level candidates only. It is not assessed by CCEA.

To use the Arrhenius equation you must be able to use several maths skills including changing the subject of an equation, substituting numerical values, manipulating units and using the 'e^x' button on your calculator.

The Arrhenius equation links the rate constant with activation energy and temperature. It is written as:

$$k = Ae^{-\frac{E_a}{RT}}$$

where:

   k is the rate constant (units depend on order of the reaction)

   A is the Arrhenius constant

   e is a mathematical constant (2.718 to 3 d.p.) (no units)

   E_a is the activation energy in $J\,mol^{-1}$

   R is the gas constant ($8.31\,J\,K^{-1}\,mol^{-1}$)

   T is the temperature measured in kelvin (K).

   $\dfrac{E_a}{RT}$ does not have any units:

$$\frac{E_a}{RT} = \frac{J\,mol^{-1}}{J\,K^{-1}\,mol^{-1} \times K}$$

Hence $e^{-\frac{E_a}{RT}}$ will not have any units either, so the Arrhenius constant will have the same units as the rate constant.

When using the Arrhenius equation be very careful about units. E_a is in $J\,mol^{-1}$ but sometimes the activation energies may be quoted in $kJ\,mol^{-1}$ and must be converted.

(A) Worked example

Calculate the rate constant to one decimal place, using the Arrhenius equation and given the following values for a reaction.

$A = 2.22 \times 10^{11}\,mol\,dm^{-3}\,s^{-1}$
$E_a = 111\,kJ\,mol^{-1}$
$T = 444\,°C$
$R = 8.31\,J\,K^{-1}\,mol^{-1}$

Step 1: ensure all values are converted to the correct units.

$A = 2.22 \times 10^{11}\,mol\,dm^{-3}\,s^{-1}$
$E_a = 111\,kJ\,mol^{-1}$. The units must be converted into $J\,mol^{-1}$ by multiplying by 1000.
$E_a = 111\,000\,J\,mol^{-1}$
$T = 444\,°C$. This must be converted to K by adding 273 to the temperature so $T = 717\,K$.
$R = 8.31\,J\,K^{-1}\,mol^{-1}$. The units are correct.

Step 2: use the equation

$$k = Ae^{-\frac{E_a}{RT}}$$

It is best to first find the value of $\dfrac{E_a}{RT}$.

$$\frac{E_a}{RT} = \frac{111\,000}{8.31 \times 717} = \frac{111\,000}{5958.27} = 18.63$$

Step 3: substitute into $k = Ae^{-\frac{E_a}{RT}}$.

$$k = 2.22 \times 10^{11}\,e^{-18.63} = 1800.7\,mol\,dm^{-3}\,s^{-1}$$

To do this on your calculator press the 'e^x' button and then type in −18.63 and then multiply by 2.22×10^{11}.

(B) Guided question

Copy out the workings and complete the answers on a separate piece of paper.

1   **A reaction has an activation energy of 115 kJ mol^{-1} at room temperature (298 K). The gas constant R is 8.31 J K^{-1} mol^{-1} and $A = 2.1 \times 10^{12}\,s^{-1}$. Calculate a value for the rate constant for the reaction at this temperature. State its units.**

Step 1: list all of the quantities given and convert them to the correct units.
$E_a = 115\,kJ\,mol^{-1}$. The units must be converted into $J\,mol^{-1}$ by multiplying by 1000;
$E_a = 115\,000\,J\,mol^{-1}$

$T = 298\,\text{K}$. The units are correct.

$R = 8.31\,\text{J K}^{-1}\,\text{mol}^{-1}$. The units are correct.

$A = 2.1 \times 10^{12}\,\text{s}^{-1}$

Step 2: use the equation:

$$k = A\text{e}^{-\frac{E_a}{RT}}$$

It is best to first find the value of $\dfrac{E_a}{RT}$.

$$\frac{E_a}{RT} = \frac{115\,000}{8.31 \times 298} = \underline{\hspace{2cm}}$$

Step 3: substitute into $k = A\text{e}^{-\frac{E_a}{RT}}$. To do this on your calculator press the 'e^x' button and then type in the value of $-\dfrac{E_a}{RT}$ and multiply by A.

The units of the rate constant are the same as the units of the Arrhenius constant.

ⓒ Practice questions

2 A reaction has an activation energy of $50\,\text{kJ mol}^{-1}$ at $20\,°\text{C}$. The gas constant R is $8.31\,\text{J K}^{-1}\,\text{mol}^{-1}$ and $A = 2.1 \times 10^{12}\,\text{s}^{-1}$. Calculate a value for the rate constant at this temperature.

3 Explain, by calculation, what happens to the rate constant if the temperature of the reaction in question 2 is increased to $30\,°\text{C}$.

Using calculators to find logs

Note: this topic is for A-level candidates only.

Large numbers are often complicated to deal with. By writing larger numbers in terms of their power to base ten, a smaller scale, called a log scale, is generated which is often easier to comprehend. In chemistry, pH is defined in terms of a logarithmic scale.

Long before calculators were invented, logarithm tables were used to simplify maths calculations. John Napier, a Scottish mathematician, published the first logarithm tables in 1641. Napier's birthplace, Merchiston Tower in Edinburgh, is now part of the facilities of Edinburgh Napier University.

Logarithms, or 'logs', express one number in terms of a base number that is raised to a power.

For example, in the expression

$$100 = 10^2$$

10 is the base and 2 is the power or index. This expression can be written in an alternative way. In terms of logs it is written:

$$\log_{10}100 = 2$$

This can be read as the 'log to base 10 of 100 is 2'. The relationship between the two expressions is shown in Figure 3.1.

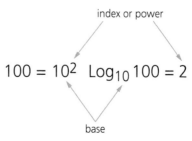

Figure 3.1 Relationship of numbers with their logs

The base can be any positive number apart from 1 but in A-level chemistry we will be using logs to base 10. Logarithms to base 10, $\log_{10}$, will be written simply as 'log' e.g. log 100 = 2.

There are two different buttons to calculate logarithms on most calculators. The 'log' button calculates logs to base ten and is the button which you should use. The 'ln' button calculates natural logs.

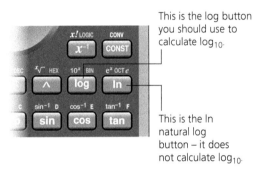

This is the log button you should use to calculate $\log_{10}$.

This is the ln natural log button – it does not calculate $\log_{10}$.

Figure 3.2 Log and ln on a calculator

Ⓐ Worked example

Check that you can use your calculator correctly by verifying the following logs:
- **log 93 is 1.97 (press log then press 93)**
- **log 0.03 is −1.52 (press log then press 0.03)**
- **log 1.1 × 10³ is 3.04 (press log, press 1.1, press exp, press 3).**

Two common equations involving logs which are used in chemistry are:

$$pH = -\log[H^+] \text{ and}$$

$$pK_a = -\log K_a.$$

Ⓐ Worked example

What is the pH of a $0.0240 \, mol \, dm^{-3}$ solution of hydrochloric acid?

Hydrochloric acid fully dissociates in solution: $HCl \rightarrow H^+ + Cl^-$ hence the concentration of H^+ ions is also $0.0240 \, mol \, dm^{-3}$.

$pH = -\log[H^+]$
$pH = -\log(0.0240)$

To calculate this on your calculator type in '−' then 'log' then '0.0240'.

$pH = 1.62$

B Guided questions

Copy out the workings and complete the answers on a separate piece of paper.

1   **If a solution has $[H^+] = 0.60\,mol\,dm^{-3}$, find the pH of the solution using the equation $pH = -\log[H^+]$.**

   $pH = -\log(0.60)$

   To calculate this on your calculator type in '−' then 'log' then '0.60'.

2   **The K_a for ethanoic acid (CH_3COOH) is $1.72 \times 10^{-5}\,mol\,dm^{-3}$. Use the equation $pK_a = -\log K_a$ to find the value pK_a.**

   $pK_a = -\log(1.72 \times 10^{-5})$

   To calculate this on your calculator type in '−' then 'log' then '1.74 exp −5'.

C Practice questions

3   Calculate the following to one decimal place.
   a $\log 0.010$
   b $\log 0.03$
   c $\log 2$
   d $\log 1.3 \times 10^{-4}$
   e $\log 2.1 \times 10^{-5}$
   f $\log 2.1 \times 10^{2}$

4   Record the value of pH to one decimal place by using the expression $pH = -\log[H^+]$ for the following values of $[H^+]$ in $mol\,dm^{-3}$.
   a   0.01
   b   0.25
   c   1.3
   d 3.1×10^{-3}
   e 3×10^{-2}

5   Calculate the pH of:
   a $0.05\,mol\,dm^{-3}$ hydrochloric acid
   b $1.0\,mol\,dm^{-3}$ sulfuric acid.

   Give your answer to one decimal place.

6   The K_a for a weak acid is 5.26×10^{-7}. Find the value of pK_a to one decimal place.

Antilogs

You must be able to use your calculator to find logs of any positive number. In addition, sometimes you may be given the log of a number and must work backwards to find the number itself. This is called finding the **antilog** of the number. An antilog is the opposite of a log. To find the antilog on most simple scientific calculators:

■   press the second function (2ndF), inverse (inv) or shift button, then,
■   press the log button; it might also be labelled the 10^x button,
■   type in the number.

An antilog 'undoes' logs by raising the base to the log number.

For example: log 100 = 2, antilog 2 = 100. In finding the antilog the calculator is performing the calculation 10^2.

Press the shift (or 2ndF) button and then the log button to access the antilog function.
The antilog function is 10^x, as shown.

Figure 3.3 How to calculate an antilog

(A) Worked examples

a   **If log x = 2.1, what is x?**

To find x you need to find the antilog of 2.1.

On your calculator type in 'shift' or 'inv' or '2ndF' followed by 'log 2.1'.

$x = 125.9$

b   **Determine the concentration, in $mol\,dm^{-3}$, of hydrochloric acid which has a pH of 0.31. Give your answer to three significant figures.**

$$pH = -log[H^+]$$

$$0.31 = -log[H^+]$$

$$-0.31 = log[H^+]$$

You need to find the antilog of -0.31. On your calculator type in '2ndF' or 'inv' or 'shift' then 'log' then '-0.31'.

$$[H^+] = 0.49$$

(B) Guided question

Copy out the workings and complete the answers on a separate piece of paper.

1   **Determine the concentration, in $mol\,dm^{-3}$, of nitric acid which has a pH of 0.75.**

Step 1: use the equation $pH = -log[H^+]$

$$0.75 = -log[H^+]$$

$$-0.75 = log[H^+]$$

Step 2: find the antilog of -0.75. On your calculator type in '2ndF' or 'inv' or 'shift' then 'log' then '-0.75'.

An alternative way of answering this is to use the equation:

$[H^+] = 10^{-pH}$

$[H^+] = 10^{-0.65}$

On your calculator this is calculated in the same way. Check that it gives the same answer.

Ⓒ Practice questions

2   Calculate the K_a of an acid which has a pK_a value of 2.99.

3   Calculate the $[H^+]$ concentration of hydrochloric acid which has
    a   pH 1.2
    b   pH 0.2
    c   pH 0.7.

4   If $\log b = -0.1$, what is b?

5   Determine the concentration, in mol dm^{-3}, of nitric acid which has a pH of 0.91. Give your answer to two decimal places.

6   Determine the concentration of sulfuric acid which has a pH of 1.1.

4 Graphs

Plotting graphs

Experiments often involve variables.

- The **independent variable** is the factor that is being changed during the experiment.
- The **dependent variable** is the factor that is being measured during the experiment.
- The **controlled variables** are the factors that are kept constant during the experiment.

Changing the independent variable causes a change in the dependent variable. After carrying out an experiment, it is often useful to plot a graph to help analyse your results. A graph is an illustration of how two variables relate to one another. In chemistry you are often asked to draw a line graph with a best-fit curve or line. A line graph is used in the situation where both variables are quantitative (numerical) and continuous (any numerical value is possible, not just whole numbers).

General construction

When plotting a graph, it is important to pay attention to the following different points:
- The independent variable is placed on the x-axis, while the dependent variable is placed on the y-axis.
- Appropriate scales should be devised for the axes. When choosing a scale you should attempt to spread the data points to use as much of the graph paper as possible — at the very least half of the graph paper should be used.

The two graphs below illustrate the importance of using appropriate scales for each axis. In the first graph the scale in the y-direction is poor, as it compresses all the points into a small section of the graph paper. Graph 2 is much better since the points fill more than half the graph grid in both the x and y directions. The origin (0,0) does not have to be included in many cases, and is not included in Graph 2.

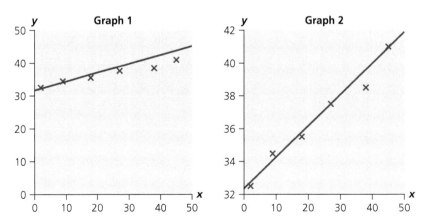

Figure 4.1 Graphs — appropriate scales

When choosing a scale it is also important to avoid using a scale which is difficult. For example, avoid using multiples of 3 or 7 or 11. Graph 3 shown below has an acceptable scale, but Graph 4 has an *x*-axis scale with difficult scale markings — it increases in threes. It is best not to use this scale.

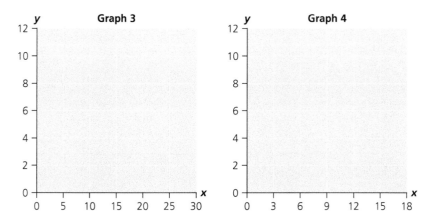

Figure 4.2 Graphs — difficult scales

It is also important when choosing a scale that you examine the data to establish whether it is necessary to start the scale(s) at zero.

- Axes should be labelled with the name of the variable followed by a solidus (/) and the unit of measurement. For example, the label may be 'temperature/°C'.
- Data points should be marked with a cross (x) so that all points can be seen when a line of best fit is drawn.
- A line of best fit should be drawn. When judging the position of the line there should be approximately the same number of data points on each side of the line. Resist the temptation to simply connect the first and last points — it may not be a best-fit line or curve. Draw a best-fit line with a thin pencil that does not obstruct the points and does not add uncertainty to the gradient.

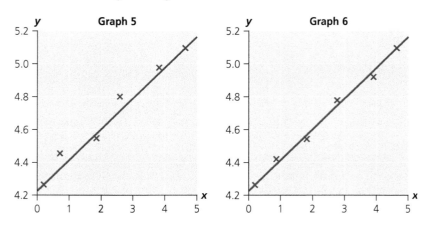

Figure 4.3 Graphs — line of best fit

Graph 5 does not show a best-fit line as there are too many points above the line. Graph 6 is a best-fit line as there are approximately the same number of data points on each side of the line.

> **TIP**
>
> A line of best fit is added by eye. You should use a transparent plastic ruler or a flexible curve to aid you.

Not all lines of best fit go through the origin. Before using the origin as a point always ask the question 'does a 0 in the independent produce a 0 in the dependent?'

When drawing a best-fit line or curve, ignore any anomalous results.

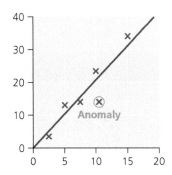

Figure 4.4 Line of best fit with anomaly

- It may be important to calculate what is happening beyond the points plotted. Extending the best-fit line or curve is a process known as **extrapolation**. Sometimes it is necessary to extrapolate to the y-axis.
- The graph should have a **title** that summarises the relationship which is being illustrated — this should include the independent variable and the dependent variable and the reaction being studied. For example, 'A graph of concentration against time for the reaction between magnesium and hydrochloric acid' or simply 'A concentration–time graph for the reaction between magnesium and hydrochloric acid'.

Worked example

In an experiment a lump of calcium carbonate was added to 50 cm³ of hydrochloric acid in a conical flask and placed on a balance. A stopwatch was started as soon as the calcium carbonate made contact with the acid, and the mass was recorded every twenty seconds. The results obtained from the experiment are shown below.

Plot a graph of mass/g (y-axis) against time/s (x-axis).

Table 4.1

Time/s	0	20	40	60	80	100	120
Mass/g	234.10	233.70	233.40	233.20	233.05	233.00	233.00

Step 1: decide on the scale for the x-axis. The graph paper has 12 squares across, hence it is appropriate to start at time zero and increase in intervals of 10, up to 120 s.

Step 2: decide on a scale for the y-axis. The y-axis has 16 squares up. The masses range from 234.10 to 233.00 which is an interval of 1.1 g. It is not sensible to start this scale at zero, as this would not spread the data points over the graph paper. Instead, the scale could increase with each square representing 0.10 g. Figure 4.5 shows a scale from 232.80 to 234.40 which spreads the data points over as much of the graph paper as possible.

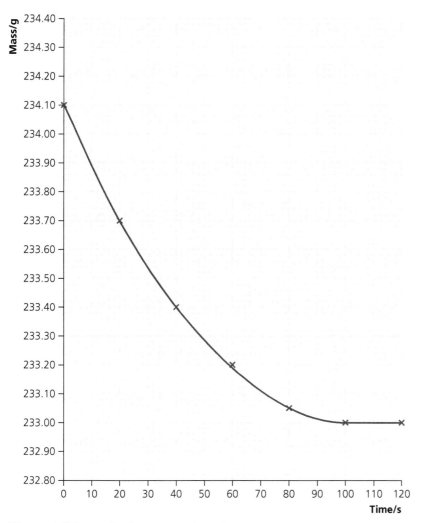

Figure 4.5 A graph of mass against time

Practice questions

1   In an experiment magnesium was reacted with sulfuric acid and the volume of
    hydrogen produced was collected and measured, every ten seconds, in a gas syringe.

    Use the data in the table to draw a graph with best-fit curve of volume of hydrogen
    against time.

    Table 4.2

Time/s	0	10	20	30	40	50	60	70	80	90	100
Volume of hydrogen/cm³	0	30	55	75	88	98	102	104	104	104	104

2   During an equilibrium reaction a gas C is produced from the reaction of gases A and B:

    $A(g) + B(g) \rightleftharpoons C(g)$

    The percentage of C in the reaction mixture varies with temperature. Plot a graph of
    percentage of C against temperature.

    Table 4.3

Temperature/°C	100	200	300	400	500
Percentage of C in equilibrium mixture/%	58	42	30	21	16

3 In an experiment some calcium carbonate and acid were placed in a conical flask on a balance and the balance reading recorded every minute. The results were recorded and the graph shown below was drawn.

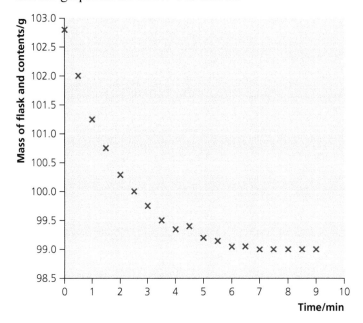

Figure 4.6

a Are there any results which you would ignore when drawing a best-fit curve?
b The *y*-axis is not labelled correctly. Write down the correct label for this axis.
c The *x*-axis is not labelled correctly. Write down the correct label for this axis.
d Suggest a title for this graph.
e Do you think the scale is appropriate in this graph? Explain your answer.

The slope and intercept of a linear graph

Every straight-line linear graph can be represented by an equation: $y = mx + c$. The coordinates of every point on the line will solve the equation if you substitute them in the equation for *x* and *y*. **The *y*-intercept is the point where the graph crosses the *y*-axis when $x = 0$**. It is the value for *y* when $x = 0$.

$$y = mx + c$$

the gradient   the *y*-intercept

Figure 4.7 The equation for a straight line

$$y = 2x + 1$$

the gradient   the *y*-intercept

Figure 4.8 The equation for the straight line on Figure 4.9

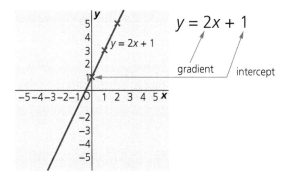

Figure 4.9 Straight-line graph

Gradient is another word for 'slope'. The higher the gradient of a graph at a point, the steeper the line is at that point. A positive gradient means the line slopes up from left to right. A negative gradient means that the line slopes downwards from left to right. For a straight-line graph the gradient is a constant value. A zero gradient graph is a horizontal line.

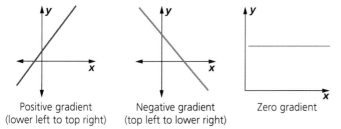

Positive gradient (lower left to top right)   Negative gradient (top left to lower right)   Zero gradient

Figure 4.10 Gradients

To find the gradient (*m*) of a straight-line graph

Choose any two points on the best-fit line, and draw a triangle with the section of best-fit line forming the hypotenuse (longest side).

In practice, a more accurate answer for the gradient is obtained when the points are as far apart as possible. In Graph 1 below, the points chosen are not far enough apart and the hypotenuse (red line) is too small to use to calculate the gradient accurately. In Graph 2, the triangle is a much better size. The hypotenuse (red line) is more than half the length of the best-fit line.

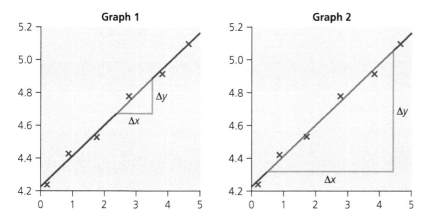

Figure 4.11 Finding a gradient

To calculate the gradient use the equation:

$$\text{gradient}\,(m) = \frac{\text{change in } y\text{-axis}}{\text{change in } x\text{-axis}} = \frac{\Delta y}{\Delta x}$$

where the numerator (top line) represents the vertical distance between the two points (the rise) and the denominator (bottom line) represents the horizontal distance between two points (the run). A simpler equation to use may be:

$$\text{gradient}\,(m) = \frac{\text{rise}}{\text{run}}$$

You should always show your working out when calculating the gradient of a line. It is also a good idea to show and label the triangle used.

(A) Worked example

Calculate the gradient of the line.

$$\text{gradient}\,(m) = \frac{\text{change in } y\text{-axis}}{\text{change in } x\text{-axis}} = \frac{\Delta y}{\Delta x}$$

$$m = \frac{2.5 - 0.5}{3.0 - 0.6} = \frac{2.0}{2.4} = 0.8$$

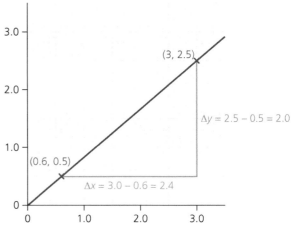

Figure 4.12

(B) Guided question

Copy out the workings and complete the answers on a separate piece of paper.

1   a Find the gradient of the lines in A, B and C.

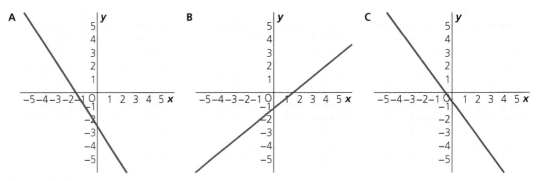

Figure 4.13

Step 1: choose two points which are far apart on the line, these will form the hypotenuse of the triangle. This step has been completed for each graph and the points are marked as dots.

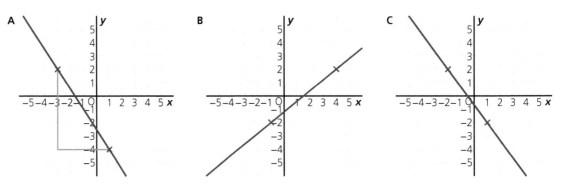

Figure 4.14

Step 2: complete the triangle (the green lines in graph A).

Step 3: find the Δy (rise) value.

Step 4: find the Δx (run) value.

Step 5: to find the gradient of each line, use the equation

$$\text{gradient } (m) = \frac{\text{change in } y\text{-axis}}{\text{change in } x\text{-axis}} = \frac{\Delta y}{\Delta x}$$

Step 6: decide if it is a positive gradient sloping from lower left to top right or a negative gradient.

b Find the *y*-intercept and write the equation for each line.

Step 1: write down the number at which the blue line cuts through the *y*-axis (at $x = 0$). This is the intercept, *c*.

Step 2: substitute the values for *m* and for *c* into the equation $y = mx + c$.

○ Practice questions

2 For each of the graphs A to G decide if the gradient is positive, negative or zero.

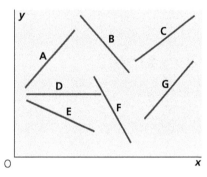

Figure 4.15

3 a Work out the gradient of each graph A to F.

b Write down the value of the *y*-intercept for each graph.

c Write the equation for each line.

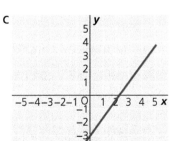

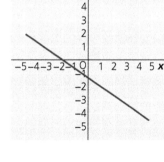

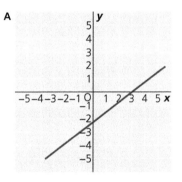

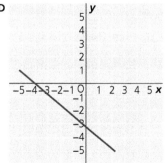

Figure 4.16

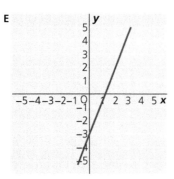

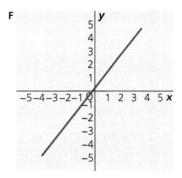

Figure 4.16 (Continued)

Units for gradient

Gradients usually have units. To work out the units, find the unit of the Δy value divided by the unit of the Δx value.

A Worked example

For the concentration against time graph shown below, work out the units of the gradient.

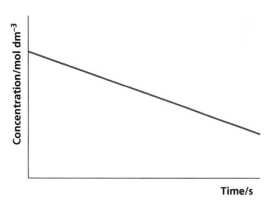

Figure 4.17

$$\text{gradient}\,(m) = \frac{\text{change in } y\text{-axis}}{\text{change in } x\text{-axis}}$$

$$= \frac{\Delta y}{\Delta x} = \frac{\text{concentration}}{\text{time}}$$

$$= \frac{\text{mol dm}^{-3}}{\text{s}} = \text{mol dm}^{-3}\,\text{s}^{-1}$$

B Guided question

Copy out the workings and complete the answers on a separate piece of paper.

1   **Work out the units of the gradient using the x and y units shown.**

Table 4.4

Units on y-axis	Units on x-axis	Units of gradient
mol dm^{-3}	min	$\dfrac{\text{mol dm}^{-3}}{\text{min}} = \text{mol dm}^{-3}\,\text{min}^{-1}$
g dm^{-3}	s	
g	cm^3	

Practice question

2 Work out the units, if any, for the gradient of a graph of:
  a concentration/mol dm^{-3} against time/s
  b rate/mol dm^{-3} s^{-1} against concentration/mol dm^{-3}
  c mass/g against time/s
  d volume/cm^3 against time/s.

Calculating the rate of change

A graph of concentration against time often
looks like that shown here. As the
concentration of the reactant falls due to it
being used up, the rate decreases and so the
curve is less steep.

For a **zero order reaction** the rate is
independent of concentration. This means
that the rate does not change as the
concentration changes over time, and so a
straight-line graph with a constant gradient
is produced.

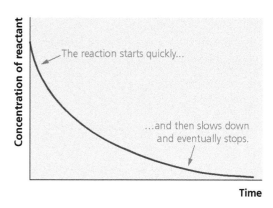

Figure 4.18 Rate of change

To calculate the rate constant of a zero order
reaction, a concentration–time graph can be plotted and the gradient gives the rate
constant.

Ⓐ Worked example

**To determine the order of a reaction between iodine and propanone in the presence of acid, a
mixture of the different solutions was prepared. Samples were removed from the mixture at
regular time intervals and titrated with sodium thiosulfate solution to find the concentration of
iodine present. A graph of the results is shown below. The gradient gives the rate.**

a **What is the order of this reaction?**

  The graph is a linear straight line with constant
  gradient (rate) showing that the rate does not
  change as the concentration of iodine changes,
  and so it is a zero order reaction.

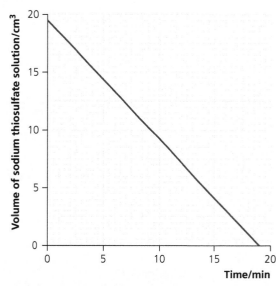

Figure 4.19

b What is the gradient of this graph and the units of the gradient?

To find the gradient choose two points far apart on the line and form a triangle as shown on the graph in green.

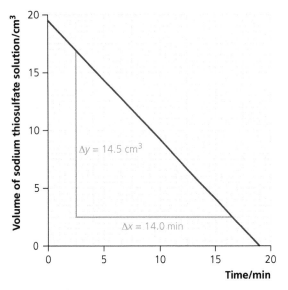

Figure 4.20

Use the equation:

$$\text{gradient}\,(m) = \frac{\text{change in } y\text{-axis}}{\text{change in } x\text{-axis}} = \frac{\Delta y}{\Delta x} = \frac{14.5}{14.0} = -1.0$$

To find the units, use the expression:

$$\text{units of gradient}\,(m) = \frac{y\text{-axis units}}{x\text{-axis units}} = \frac{\text{cm}^3}{\text{s}} = \text{cm}^3\text{s}^{-1}$$

B Guided question

Copy out the workings and complete the answers on a separate piece of paper.

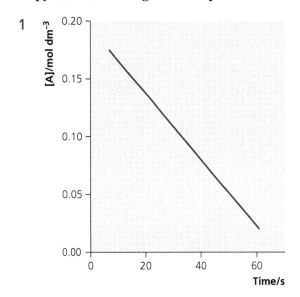

Figure 4.21

Use the concentration–time graph to:

a determine the order of the reaction

This is a linear straight-line graph, hence the gradient (rate) is constant and does not change as concentration changes. It is a zero order reaction and the gradient of the graph represents the rate.

b calculate the rate constant and state its units.

Step 1: to find the gradient, choose two points far apart on the line and form a triangle as shown on the graph in green.

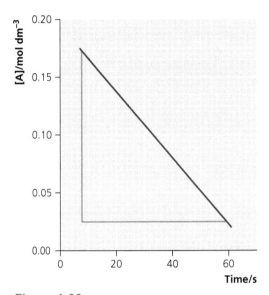

Figure 4.22

Step 2: use the equation

$$\text{gradient}\,(m) = \frac{\text{change in } y\text{-axis}}{\text{change in } x\text{-axis}} = \frac{\Delta y}{\Delta x}$$

Step 3: to find units, use the expression

$$\text{units of gradient} = \frac{y\text{-axis units}}{x\text{-axis units}} = \frac{\text{mol dm}^{-3}}{\text{s}} = \underline{\hspace{3cm}}$$

The gradient is the rate constant.

TIP

In chemistry it is often necessary to draw concentration–time graphs and rate–concentration graphs. Rate numbers are often very small and can be difficult to plot. For example, rate = $0.0015\,\text{mol dm}^{-3}\,\text{s}^{-1}$ or rate = $0.0024\,\text{mol dm}^{-3}\,\text{s}^{-1}$. It can be easier to convert the numbers into standard form e.g. $0.00015 = 1.5 \times 10^{-4}\,\text{mol dm}^{-3}\,\text{s}^{-1}$ and $0.0024 = 2.4 \times 10^{-4}\,\text{mol dm}^{-3}\,\text{s}^{-1}$ and then plot the numbers 1.5 and 2.4, labelling the axis in the form 'rate $\times\ 10^{-4}/\text{mol dm}^{-3}\,\text{s}^{-1}$'.

Practice questions

2   For the reaction $X + Y \rightarrow Z$, the results from an experiment are shown in Table 4.5.

Table 4.5

[X]/mol dm^{-3}	Rate/mol dm^{-3} s^{-1}
0.01	0.0025
0.02	0.0050
0.03	0.0075
0.04	0.0100
0.05	0.0125

Plot a graph of rate against concentration and determine the gradient.

3   The graph below shows the concentration of iodine against time.

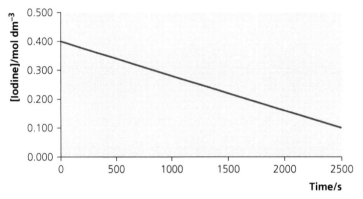

Figure 4.23 Concentration of iodine against time

a   What is the order of the reaction with respect to iodine?

b   Use the graph to calculate the rate of the reaction.

Tangents and measuring the rate of change

Tangents to a curve

The word tangent means 'touching' in Latin. The tangent is a straight line which just touches the curve at a given point and does not cross the curve.

To draw a tangent at a point (x, y), as shown in Figure 4.24:

■   place your ruler through the point (x, y)

■   make sure your ruler goes through the point and does not touch the curve at any other point

■   draw a ruled pencil line passing through point (x, y).

To calculate the gradient of a curve at a particular point it is necessary to draw a tangent to the curve at the point and calculate the gradient of the tangent.

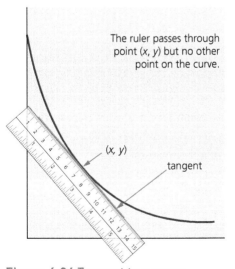

The ruler passes through point (x, y) but no other point on the curve.

(x, y)

tangent

Figure 4.24 Tangent to a curve

 Worked example

In an experiment a student recorded the total volume of gas collected in a reaction at 20-second intervals.

Table 4.6

Time/s	0	20	40	60	80	100	120
Volume of gas/cm³	0	21	42	56	65	72	72

a   **Plot a graph using the data shown and draw a line of best fit.**

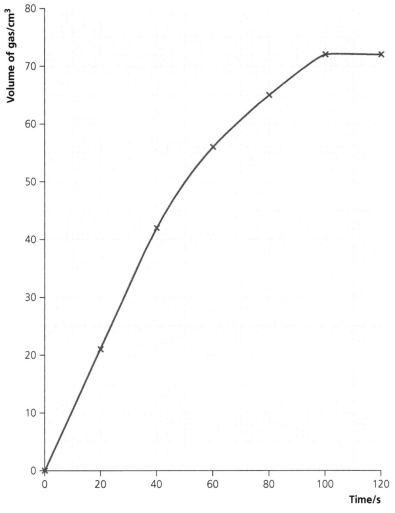

Figure 4.25

The graph will be a graph of volume of gas (y-axis) against time (x-axis). Allocating one large square for 20 s is a scale which is suitable on the x-axis, and one large square for 10 cm³ of gas is suitable on the y-axis.

b   **Use the graph to calculate the rate of reaction at 60 s. State the units.**

The graph is a curve and to find the rate of reaction at 60 s, a tangent to the curve must be drawn at 60 s, as shown on Figure 4.26 in red.

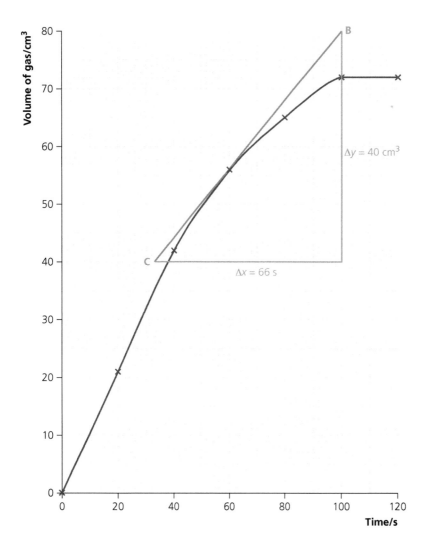

Figure 4.26

The rate is given by the gradient of this tangent. To find the gradient choose two points, B and C, far apart on the line and form a triangle as shown in green on Figure 4.26.

$$\text{gradient}\,(m) = \frac{\text{change in } y\text{-axis}}{\text{change in } x\text{-axis}} = \frac{\Delta y}{\Delta x} = \frac{40}{66} = 0.61\,(\text{accurate to 2 s.f.})$$

$$\text{units} = \frac{\text{cm}^3}{\text{s}} = \text{cm}^3\,\text{s}^{-1}$$

> **TIP**
>
> In exam questions, data given may include an anomalous result. Make sure you ignore this result when drawing a best-fit line.

B Guided question

Copy out the workings and complete the answers on a separate piece of paper.

1

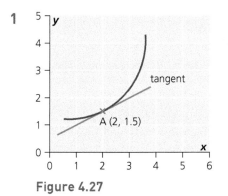

Figure 4.27

What is the gradient of the curve at point A?

Step 1: to find the gradient of the curve at point A, a tangent to the curve at point A must be drawn. It is shown in red.

Step 2: choose two points, B and C, far apart on the line and form a triangle as shown in Figure 4.28.

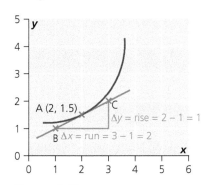

Figure 4.28

Step 3: calculate the gradient using:

$$\text{gradient}\,(m) = \frac{\text{change in } y\text{-axis}}{\text{change in } x\text{-axis}} = \frac{\Delta y}{\Delta x}$$

To find the slope of the curve at any other point, you would need to draw a tangent line at that point and then determine the slope of that tangent line. This is the method used to find the order of a reaction using the initial rates method and is illustrated in Figure 4.29.

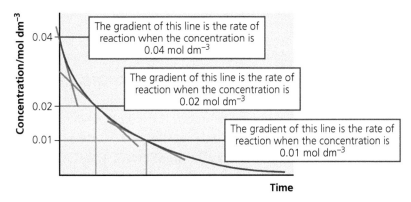

Figure 4.29

2 Calculate the rate of reaction when the concentration of ester is 0.200 mol dm⁻³.

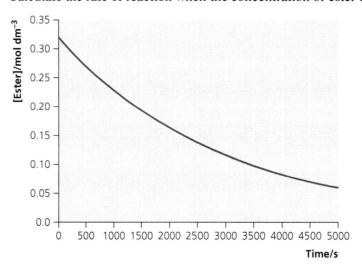

Figure 4.30

3 The volume of carbon dioxide gas produced over time, when calcium carbonate reacted with hydrochloric acid, was recorded in the table below.

Table 4.7

Time/s	0	10	20	30	40	50	60	70	80	90	100
Volume of carbon dioxide/cm³	0	22	35	43	48	52	55	57	58	58	58

a Plot a graph of the volume of carbon dioxide against time.
b Calculate the rate of reaction at 20 seconds by drawing a tangent to the curve.
c Calculate the rate of reaction at 60 seconds by drawing a tangent to the curve.

Logarithmic scales

Logs are a powerful way to reduce a set of numbers that range over many orders of magnitude to a smaller more manageable scale. pH is a logarithmic scale. The table below shows how taking the log of the hydrogen ion concentration reduces the range of numbers being dealt with. If values for [H⁺] range from 0.1 to 0.00000000000010, the pH can fit in the range 1 to 14!

Table 4.8

pH	0	1	7	13	14
[H⁺]/mol dm⁻³	1	10^{-1}	10^{-7}	10^{-13}	10^{-14}

Presentation of data on a logarithmic scale can be helpful when drawing graphs. Logarithmic scales are used when the data:

■ covers a large range of values — the use of logs reduces a wide range to a more manageable size which is easier to plot
■ may contain exponential or power laws since these will show up as straight lines. You could be asked to draw graphs using logs or natural logs for rate equations and for the Arrhenius equation. You do not have to work out the log expression, this is generally given in the question. Rate graphs and Arrhenius equation graphs are important logarithmic graphs.

Rate graphs

For a second order reaction, rate $= k[A]^2$, a graph of rate against concentration would give a curve. If logs are taken of both sides, however, this gives:

log rate $=$ log $k + 2$log[A]

$y = c + mx$

TIP

Note that log $x^2 = 2$log x,
so log[A]$^2 = 2$log[A].

This has the same form as a straight-line graph $y = mx + c$. Consequently, a plot of log rate against log[A] gives a straight-line graph whose intercept is the value for log k and the gradient is equal to the order of the reaction. This treatment is valid for any order values. The gradient (slope) of the graph line is always equal to the order.

Sometimes you may be asked to plot a graph by first calculating the log of some values and then plotting the logs. Simply calculate the log values on your calculator and record them before plotting.

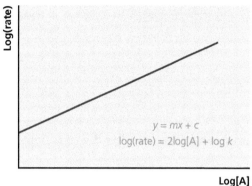

Figure 4.31 Gradient from a rate graph

Arrhenius equation

By taking natural logs of both sides, the Arrhenius equation, $k = Ae^{-\frac{E_a}{RT}}$ can be written in a more useful form:

$$\ln k = \ln A - \frac{E_a}{RT}$$

This natural log form is often given in questions. $\frac{E_a}{RT}$ is the inverse of the exponential $e^{-\frac{E_a}{RT}}$. This form is more useful because it follows the mathematical formula for a straight line, $y = mx + c$.

i.e. $\ln k = -\frac{E_a}{RT} + \ln A$

and so:

$y = \ln k$

$x = \frac{1}{T}$

$c = \ln A$

$m = -\frac{E_a}{R}$

If a graph of ln k against $\frac{1}{T}$ is drawn, the gradient can be used to calculate E_a and the y-intercept is ln A.

Worked example

In an experiment to determine the activation energy for the reaction between zinc and hydrochloric acid, the data shown in the table below was obtained.

Table 4.9

Temperature/K	$\frac{1}{T} \times 10^{-3}/K^{-1}$	ln k
283	3.53	−4.80
299	3.34	−3.56
311	3.22	−3.00
322	3.11	−2.25
329	3.04	−1.79

The activation energy can be found using the equation:

$$\ln k = -\frac{E_a}{RT} + \ln A$$

Use the data in the table to plot a graph of ln k against $\frac{1}{T}$ and use it to determine the activation energy.

Step 1: the x-axis values stretch from 3.04×10^{-3} to 3.53×10^{-3}, so a sensible scale to use is from 3.00 to 3.60 with each large square representing $0.10\,K^{-1}$.

Step 2: the y-axis values stretch from −1.79 to −4.8, so a suitable scale is to allow each large square to represent 1.0.

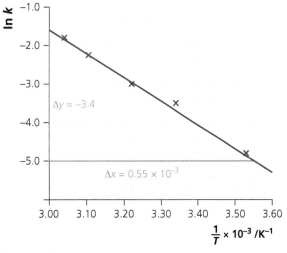

Figure 4.32

Step 3: to calculate the gradient, draw in a triangle (green lines) and use the equation:

$$\text{gradient}\,(m) = \frac{\Delta y}{\Delta x} = \frac{-3.4}{0.55 \times 10^{-3}} = -6181.82$$

$$\text{gradient}\,(m) = -6181.82 = -\frac{E_a}{R}$$

The value of R, the gas constant, is 8.31.

$$-6181.82 = -\frac{E_a}{8.31}$$

Step 4: rearrange this equation by multiplying each side by 8.31.

$$-6181.82 \times 8.31 = -\frac{E_a}{8.31} \times 8.31$$

$$-51\,370.92 = -E_a$$

$$E_a = 51\,370.92\,\text{J}\,\text{mol}^{-1} = 51.4\,\text{kJ}\,\text{mol}^{-1}$$

B Guided question

Copy out the workings and complete the answers on a separate piece of paper.

1 **Determine the activation energy, in kJ mol⁻¹, and the value of A, given the following data:**

Table 4.10

T/K	278	288	298	308	318
$k/\text{mol}^{-1}\,\text{dm}^3\,\text{s}^{-1}$	0.013	0.530	1.04	1.67	2.14
$\ln k$	−4.34				
$1/T\;K^{-1}$	3.6×10^{-3}				

To determine the activation energy, a graph of $\ln k$ against $\frac{1}{T}$ should be drawn.

Step 1: check that temperature is in the correct units (K) and then use your calculator to calculate $\frac{1}{T}$.

Step 2: calculate $\ln k$.

> **TIP**
>
> Remember to use the 'ln' button on your calculator not the 'log' button.

Step 3: plot a graph with $\ln k$ on the y-axis and $\frac{1}{T}$ on the x-axis. It may be easiest to plot '$\frac{1}{T} \times 10^{-3}$' on the x-axis.

Step 4: find the gradient of this straight-line graph.

$$\text{gradient}\,(m) = -\frac{E_a}{R}$$

Step 5: rearrange this equation, $mR = -E_a$

$$E_a = -mR$$

Step 6: multiply the gradient by −8.31 to find the activation energy in $J\,mol^{-1}$.

Step 7: divide the value by 1000 to give the activation energy in $kJ\,mol^{-1}$.

Step 8: extrapolate your graph back to the y-axis and read off the value. This is the value of ln A.

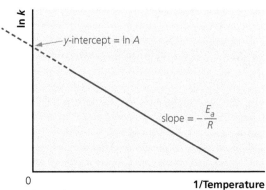

Figure 4.33

Step 9: ln A = y-intercept, then to find A:

$$e^{y\text{-intercept}} = A$$

To do this on your calculator, press the 'e^x' button and type in the intercept value.

ⓒ Practice questions

2 Determine the activation energy, in $kJ\,mol^{-1}$, given the following data:

Table 4.11

$T/°C$	327	427	527	627	727
k/s^{-1}	0.00034	0.0031	0.029	0.100	0.380

3 Determine the activation energy, in $kJ\,mol^{-1}$, given the following data:

Table 4.12

$T/°C$	55	85	125	205	305
$k/mol^{-1}\,dm^3\,s^{-1}$	5.7×10^{-8}	1.6×10^{-4}	3.9×10^{-2}	7.8×10^{-1}	6.3

5 Geometry and trigonometry

Representing shapes of molecules in 2D and 3D form

Table 5.1 shows the difference between 2D and 3D shapes.

Table 5.1

3D shapes	2D shapes
Have three dimensions — height, depth and width.	Have two dimensions — length and width. They have no depth and are flat.
These figures can be drawn on a sheet of paper using wedged and dashed lines.	These figures can be drawn on a sheet of paper in one plane using solid lines.
3D figures deal with three coordinates: x-coordinate, y-coordinate and z-coordinate.	2D figures deal with two coordinates: x-coordinate and y-coordinate.

Displayed formulae and structural formulae represent molecules in 2D with all the bonds shown as solid lines. No information about the orientation or shape of the molecules is given. The displayed formulae of methane and ethanol are shown below.

H—C—H   H—C—C—O—H

Figure 5.1 Methane and ethanol

The benefits of drawing 3D molecules include the ability to show information on bond angles, shapes and isomers, including stereoisomers.

Symbols used to draw molecules in 3D are shown in Table 5.2.

Table 5.2

Type of bond	Orientation
Normal bond ———	Bond lies in the plane of the paper
Dashed bond ---------	Bond extends backwards, effectively into the page
Wedged bond ◣■■■	Bond extends forwards, effectively out of the page

A 3D structural representation of methane CH_4, with its ball and stick model, is shown in Figure 5.2.

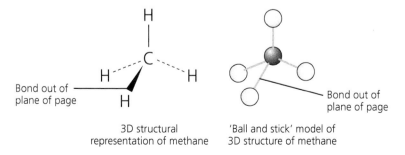

3D structural
representation of methane

'Ball and stick' model of
3D structure of methane

Figure 5.2 Methane — 3D structural representation and 'ball and stick' model

You will need to draw 3D representations of molecules, when you are asked to draw the shape of a molecule, as determined using valence shell electron pair repulsion principles. Some examples are shown in Table 5.3.

Table 5.3 The 3D shapes of some molecules

3D shape	Name of shape	Number of lone pairs	Number of bonded pairs
$Cl\text{—}Be\text{→}Cl$ 180°	linear	0	2
B with three F, 120°	trigonal planar	0	3
C with four H, 109°	tetrahedral	0	4
P with five F, 90°, 120°	trigonal bipyramidal	0	5
S with six F, 90°	octahedral	0	6
N with three H, 107°	pyramidal	1	3
O with two H, 104.5°	bent	2	2
Br with four F	square planar	2	4

Before drawing the shape of a molecule or ion you need to work out the number and type of electron pairs on the central atom (the one the other atoms are bonded to). Use the following method.

1 Use your Periodic Table to find the number of outer shell electrons around the central atom.
2 Add one if it is a negative ion, subtract one if it is a positive ion.
3 The formula will tell you how many atoms are bonded to the central atom. Each will share one of their atoms. Add one for each atom shared.
4 You now have the total number of electrons around the central atom. Divide this number by two to find the total number of electron pairs.
5 Take away the number of bonded atoms to find the number of lone pairs.

Then refer to Table 5.3 and draw the shape.

A Worked example

What is the shape of CF_4?

■ Carbon, the central atom, has an electronic configuration of 2,4 and 4 electrons in its outer shell.
■ There are 4 fluorine atoms bonded to carbon, so there are 4 electrons coming from the fluorine atoms giving a total of 8 electrons.
■ 8 electrons = 4 electron pairs
■ 4 electron pairs – 4 bonded atoms = 0 lone pairs
■ In CF_4 there are 4 bonding pairs. By referring to Table 5.3, you can conclude that this molecule has a tetrahedral shape.

B Guided question

1 **What is the shape of BrF_4^-?**

Bromine has 7 electrons in its outer shell.

There is a −1 charge on the ion, so add one to give 8 electrons.

There are 4 fluorine atoms bonded, so there are 4 electrons coming from the fluorine atoms giving a total of ___ electrons.

_____ electrons = _____ electron pairs

_____ electron pairs – _____ bonded atoms = _____ lone pairs

In BrF_4^- there are _____ bonding pairs and ___ lone pairs and so the shape is _____.

C Practice questions

2 What is the shape of each of the following molecules?
  a SiH_4 b PH_3 c BrF_2^+
  d PH_4^+ e $TlBr_3^{2-}$ f H_3O^+

3 Draw the shape of
  a NH_4^+ b BF_4^- c H_3O^+.

Understand the symmetry of 2D and 3D shapes

Optical isomers contain a **chiral** or asymmetric centre, which is a central carbon atom that has four different atoms or groups attached to it. You must be able to recognise a chiral centre from 2D or 3D representations.

The molecule 2-chlorobutane is shown in Figure 5.3. The four different groups attached to a carbon atom are circled and the chiral centre is marked with an asterisk *.

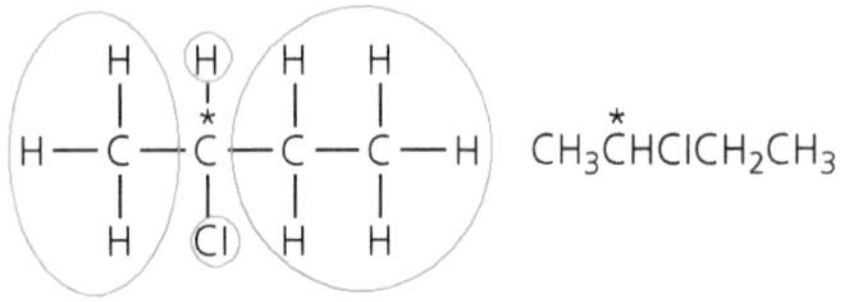

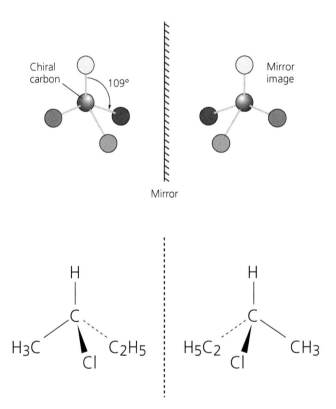

Figure 5.3 2-chlorobutane

Compounds with a chiral centre are tetrahedral and asymmetric, i.e. they have no centre, plane or axis of symmetry. As a result two tetrahedral arrangements occur in space. One is the mirror image of the other and they cannot be superimposed on each other as shown in Figure 5.4. The two mirror images are optical isomers.

Figure 5.4 Optical isomers

A Worked example

Lactic acid, $CH_3CH(OH)COOH$, is optically active. Label the asymmetric centre with a * and draw the two optical isomers.

■ The central carbon atom has four different groups bonded to a single carbon atom (H, CH_3, OH and COOH).

■ The chiral centre (asymmetric centre) is labelled with *.

$CH_3C*H(OH)COOH$

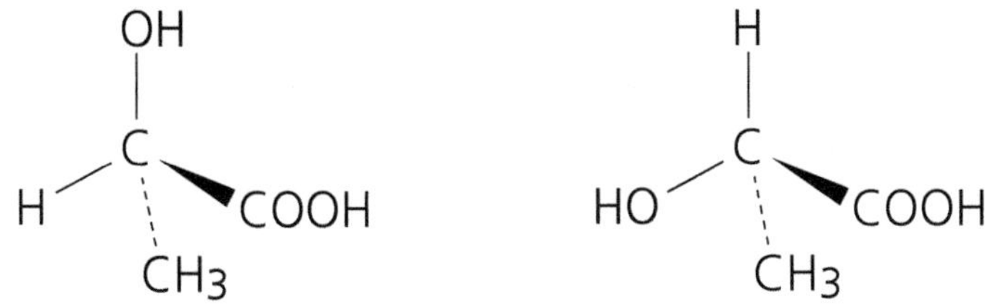

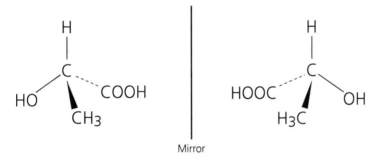

Figure 5.5 Optical isomers of lactic acid

These are the two optical isomers. They are shown simply by exchanging two of the groups bonded to the asymmetric centre. It does not matter which two groups are swapped. Alternatively, a mirror image can be drawn.

Figure 5.6 Optical isomers of lactic acid

B Guided question

Copy out the workings and complete the answers on a separate piece of paper.

1   **How many chiral centres are present in the molecule shown?**

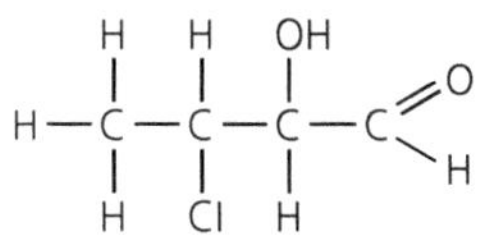

Figure 5.7

■ Look at the individual carbon atoms. The CHO group has a carbon atom but it does not have four different groups bonded — there are only three groups attached. It is not a chiral centre.

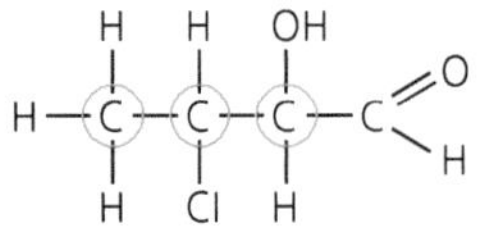

Figure 5.8

- The three circled carbon atoms have four groups bonded to them.

  Step 1: what atoms or groups are bonded to the first circled carbon atom?

- Three H atoms and CHClCHOHCHO, so it is not a chiral centre as there are not four different atoms or groups attached.

  Step 2: what atoms or groups are bonded to the second circled carbon atom?

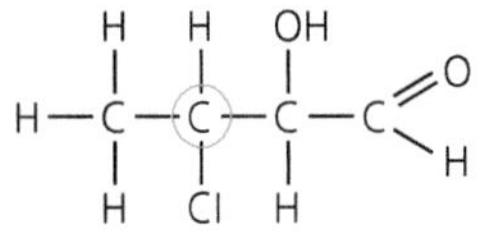

Figure 5.9

Step 3: what atoms are bonded to the third circled carbon atom?

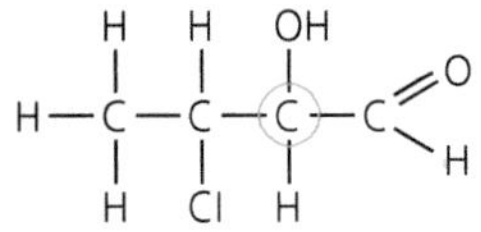

Figure 5.10

Step 4: there are _____ chiral centres.

C Practice question

2   Decide if the following molecules have chiral centres and if so, draw the 3D structures of the optical isomers.
   a NH_2CHCH_3COOH
   b NH_2CH_2COOH
   c $CH_3CH_2CH_2OH$
   d $NH_2CHClCOOH$

Exam-style questions

 Worked examples

Most chemistry AS and A-level exam questions incorporate several of the maths skills explained in this book. The following worked examples each involve several different skills. These are stated at the start of the answer and each question is worked through in detail. All exam boards use a mixture of structured questions and multiple choice questions.

a **2.00 tonnes of ammonia were reacted with carbon dioxide in the industrial production of 2.35 tonnes of urea, NH_2CONH_2.**

$$2NH_3 + CO_2 \rightarrow NH_2CONH_2 + H_2O$$

Calculate the percentage yield of urea to two decimal places.

To answer this question you must understand the following maths skills:
- converting between units (page 7)
- substituting numerical values into equations (page 53)
- using standard form (page 25)
- using ratios (page 29)
- changing the subject of an equation (page 50)
- rounding to decimal places (page 20).

Percentage yield is calculated using the equation:

$$\text{percentage yield} = \frac{\text{actual yield} \times 100}{\text{theoretical yield}}$$

The theoretical yield must be calculated first, using the mass of ammonia to calculate the number of moles of ammonia. By ratio the number of moles of urea, and hence the mass, can be calculated.

Step 1: use the equation.

$$\text{moles of ammonia} = \frac{\text{mass of ammonia in g}}{M_r}$$

The mass in the question is given in tonnes. To convert from tonnes to grams multiply by 10^6.

Therefore the mass of ammonia is 2.00×10^6 g.

To find M_r add the A_r of N and the A_r of 3H = 14.0 + (3 × 1.0) = 17.0.

Therefore:

$$\text{moles of ammonia} = \frac{2.00 \times 10^6}{17.0} = 117\,647.0588$$

Step 2: from the chemical equation, the ratio of moles of ammonia to moles of urea is $2:1$. Hence there are twice as many moles of ammonia as urea, and to find the moles of urea, divide the moles of ammonia by two.

$$\frac{117\,647.0588}{2} = 58823.5294$$

Step 3:

$$\text{moles of urea} = 58823.5294 = \frac{\text{mass of urea in g}}{M_r}$$

M_r urea, $NH_2CONH_2 = 2N + 4H + 1C + 1O$

$$= (2 \times 14.0) + (4 \times 1.0) + (1 \times 12.0) + (1 \times 16.0) = 60.0$$

$$58823.5294 = \frac{\text{mass of urea in g}}{60.0}$$

To find the mass of urea, you must change the subject of the equation. First switch sides.

$$\frac{\text{mass of urea in g}}{60.0} = 58823.5294$$

You require the mass of urea on its own, so multiply both sides of the equation by 60.0 and simplify.

$$\frac{\text{mass of urea in g}}{60.0} \times 60.0 = 58823.5294 \times 60.0$$

Mass of urea in g $= 3\,529\,411.764\,g$

To convert this to tonnes, divide by 10^6.

$$\frac{3529\,411.764}{10^6} = 3.53$$

This is the theoretical yield of urea.

Step 4: substitute the values of theoretical yield and the actual yield given in the question into the equation:

$$\text{percentage yield} = \frac{\text{actual yield} \times 100}{\text{theoretical yield}}$$

$$= \frac{2.35}{3.53} \times 100 = 66.5722$$

Step 5: to correct this to two decimal places, underline the numbers up to two after the point.

66.5722

The next number is two and so the rounding rule to use is 'if the number is four or less, do not round up'.

Therefore the answer is 66.57% (2 d.p.)

b   **3.56 g of calcium chloride was dissolved in water and reacted with an excess of sulfuric acid to form a precipitate of insoluble calcium sulfate. If the percentage yield of calcium sulfate was 83.4%, calculate the mass of calcium sulfate precipitate formed. Give your answer to an appropriate number of significant figures.**

$$CaCl_2 + H_2SO_4 \rightarrow CaSO_4 + 2HCl$$

To answer this question you must understand the following maths skills:

- substituting numerical values into equations (page 53)
- using ratios (page 29)
- changing the subject of an equation (page 50)
- rounding to the appropriate number of significant figures (page 37).

Percentage yield is calculated using the equation:

$$\text{percentage yield} = \frac{\text{actual yield} \times 100}{\text{theoretical yield}}$$

Step 1: the theoretical yield must be found first. Find the amount in moles of calcium chloride.

$$\text{amount in moles} = \frac{\text{mass of calcium chloride in g}}{M_r}$$

$$M_r = (1 \times 40.1) + (2 \times 35.5) = 111.1$$

$$\text{amount in moles} = \frac{3.56}{111.1} = 0.03204$$

The equation has a $1:1$ ratio so the amount in moles of calcium sulfate = 0.03204. The M_r of calcium sulfate is $(1 \times 40.1) + (1 \times 32.1) + (4 \times 16.0) = 136.2$.

$$\text{amount in moles} = \frac{\text{mass of calcium sulfate in g}}{M_r}$$

$$0.03204 = \frac{\text{mass of calcium sulfate in g}}{136.2}$$

To find the mass of calcium sulfate, you must change the subject of the equation. First switch sides.

$$\frac{\text{mass of calcium sulfate in g}}{136.2} = 0.03204$$

You require the mass of calcium sulfate on its own, so multiply both sides of the equation by 136.2 and simplify.

$$\frac{\text{mass of calcium sulfate in g}}{\cancel{136.2}} \times \cancel{136.2} = 0.03204 \times 136.2$$

Therefore the theoretical mass of calcium sulfate in g = 4.3638 g

Step 3: substitute the values into the equation.

$$\text{percentage yield} = \frac{\text{actual yield} \times 100}{\text{theoretical yield}}$$

$$83.4 = \frac{\text{actual yield} \times 100}{4.3638}$$

To find the actual yield of calcium sulfate you need to change the subject of the equation. First switch sides.

$$\frac{\text{actual yield} \times 100}{4.3638} = 83.4$$

To get the actual yield on its own, you need to multiply both sides of the equation by 4.3638 and divide each side by 100.

$$\frac{\text{actual yield} \times \cancel{100} \times \cancel{4.3638}}{\cancel{4.3638} \times \cancel{100}} = \frac{83.4 \times 4.3638}{100}$$

actual yield = 3.639409 g

Step 4: to find the appropriate number of significant figures to use in your answer, write down the number of significant figures in the data given.

3.56 g = 3 s.f.
83.4% = 3 s.f.

The answer can be given with accuracy to three significant figures.

Step 5: to round 3.639409 g to three significant figures, underline the first three figures:

3.639409

Look at the next number: it is 9 and greater than 5, so should be rounded up.
The answer is 3.64 g to 3 s.f.

c **8.10 g of a solid mixture was weighed out and dissolved in water. The solution was made up to 1 dm³. To find the percentage of sodium carbonate in this mixture, 25.0 cm³ of the solution was titrated with hydrochloric acid of concentration 0.100 mol dm⁻³.**

$$Na_2CO_3 + 2HCl \rightarrow 2NaCl + CO_2 + H_2O$$

The experiment was repeated and the results shown in the table below.

Table 6.1

Titration	1	2	3	4
Final burette reading/cm³	23.60	24.10	23.90	23.50
Initial burette reading/cm³	0.50	0.15	0.70	0.35
Titre/cm³	23.10	23.95	23.20	23.15

Calculate the mass of sodium carbonate present in the mixture, and calculate the percentage of sodium carbonate in the mixture.

To answer this question you must understand the following maths skills:
- calculating averages (page 42)
- using ratios (page 29)
- changing the subject of an equation (page 50)
- substituting numerical values into an equation (page 53).

Step 1: the average titre must be calculated using only concordant titres. A concordant titre is obtained when the titres are within ±0.10 cm³ of each other, hence the value 23.95 is not used.

To calculate the average, add the values together and divide by the total number of values (3).

23.10 + 23.20 + 23.15 = 69.45

$$\frac{69.45}{3} = 23.15$$

Step 2: calculate the number of moles of hydrochloric acid by substituting values into the equation:

$$n = \frac{v \times c}{1000}$$

$$= \frac{23.15 \times 0.100}{1000}$$

$$= 0.002315$$

Step 3: from the balanced symbol equation, the ratio is 2 moles of hydrochloric acid to 1 mole of sodium carbonate.

Hence to find the number of moles of sodium carbonate, divide the moles of hydrochloric acid by 2.

$$\frac{0.002315}{2} = 0.0011575$$

Step 4: use the equation:

$$n = \frac{v \times c}{1000}$$

$$0.0011575 = \frac{25.0 \times c}{1000}$$

You need c as the subject of the equation so switch sides.

$$\frac{25.0 \times c}{1000} = 0.0011575$$

To get c on its own on the left of the equation you need to multiply both sides by 1000 and divide both sides by 25.0.

$$\frac{25.0 \times c \times 1000}{1000 \times 25.0} = \frac{0.0011575 \times 1000}{25.0}$$

$$c = 0.0463 \, \text{mol} \, \text{dm}^{-3}$$

Step 5: use the equation:

$$\text{mol} \, \text{dm}^{-3} = \frac{\text{g} \, \text{dm}^{-3}}{M_r}$$

M_r of Na_2CO_3 is $(2 \times 23.0) + (1 \times 12.0) + (3 \times 16.0) = 106.0$

$$\text{mol} \, \text{dm}^{-3} = \frac{\text{g} \, \text{dm}^{-3}}{M_r}$$

$$0.0463 = \frac{\text{g} \, \text{dm}^{-3}}{106.0}$$

To change the subject of the equation switch sides and multiply both sides by 106.0 to get $\text{g} \, \text{dm}^{-3}$ on its own on one side of the equation.

$$\frac{\text{g} \, \text{dm}^{-3} \times \cancel{106.0}}{\cancel{106.0}} = 0.0463 \times 106.0 = 4.9078 \, \text{g} \, \text{dm}^{-3}$$

Step 6: the original mixture was $8.10 \, \text{g} \, \text{dm}^{-3}$ and you have worked out that it contains $4.9078 \, \text{g} \, \text{dm}^{-3}$ sodium carbonate.

The percentage of the mixture which is sodium carbonate is:

$$\frac{4.9078 \times 100}{8.10} = 60.6\% \text{ to 3 s.f.}$$

This is given as three significant figures as the data with least accuracy used in the calculation has three significant figures.

Note: example d is for A-level candidates only.

d **Ammonium chloride can dissociate to form ammonia and hydrogen chloride.**

$$NH_4Cl \rightleftharpoons NH_3 + HCl$$

At 298 K, $\Delta H = +176\,kJ\,mol^{-1}$ and $\Delta G = +91.20\,kJ\,mol^{-1}$. Calculate ΔG for this reaction at 950 K and determine if the reaction occurs spontaneously at this temperature. Give your answer to an appropriate number of significant figures.

To answer this question you must understand the following maths skills:

- substituting numerical values into equations (page 53)
- changing the subject of an equation (page 50)
- rounding to the appropriate number of significant figures (page 37).

Step 1: the question gives information about ΔH, ΔG and a temperature in kelvin.

Use the equation $\Delta G = \Delta H - T\Delta S$ and substitute the values given.

$$91.2 = 176 - 298\Delta S$$

Step 2: to calculate ΔS you need to make it the subject of the equation. First, switch sides so that ΔS is on the left of the equation.

$$176 - 298\Delta S = 91.2$$

You require the ΔS by itself as the subject on the left-hand side so you need to subtract 176 from the left side but, to keep the equation true, you need to subtract 176 from the right side as well, and then simplify.

$$176 - 298\Delta S - 176 = 91.2 - 176$$

$$-298\Delta S = -84.8$$

You require ΔS by itself as the subject so you need to divide the left side by -298 and, to keep the equation true, you must also divide the right side by -298.

$$\frac{-298\Delta S}{-298} = \frac{-84.8}{-298}$$

$$\Delta S = 0.284564\,kJ\,K^{-1}\,mol^{-1}$$

Step 3: determine the number of significant figures in each part of the data given.

Table 6.2

Data	Number of significant figures
T	3
ΔS	3
ΔH	3

The measurement which has the least number of significant figures is the least accurate. It has three significant figures, hence your answer must be given to three significant figures.

Step 4: to write the number to three significant figures underline the first three non-zero numbers.

0.<u>284</u>564

Look at the next number: it is 4, so use the rounding rule 'if it is less than 4 do not round up'.

$\Delta S = 0.285\,\text{kJ}\,\text{mol}^{-1}$

Step 5: to find ΔG at the different temperature of 950 K, once again use the equation $\Delta G = \Delta H - T\Delta S$ and substitute the values.

$\Delta G = 176 - 950 \times 0.285$

$= 176 - 270.75 = -94.75\,\text{kJ}\,\text{mol}^{-1}$

Again, the measurement which has the least number of significant figures is the least accurate. All data had three significant figures. Hence your answer must be given to three significant figures.

Step 6: to write the number to three significant figures underline the first three non-zero numbers.

$-\underline{94.7}5$

Look at the next number, and use the rule 'if the number is greater than or equal to five, round up'.

The answer, to three significant figures, is $\Delta G = -94.8\,\text{kJ}\,\text{mol}^{-1}$

The reaction takes place spontaneously at 950 K because the value of ΔG is less than zero.

Exam-style questions

AS and A-level questions

1 Iron is extracted from iron(III) oxide according to the equation:

$Fe_2O_3(s) + 3CO(g) \rightarrow 2Fe(l) + 3CO_2(g)$

Calculate the atom economy of extracting iron in this process. **(3)**

2 Hydrazine (N_2H_4) is used for rocket fuel. Calculate the atom economy for hydrazine production. **(3)**

$2NH_3 + NaOCl \rightarrow N_2H_4 + NaCl + H_2O$

3 A sample of potassium permanganate, $KMnO_4$, was heated to constant mass in a crucible and the following reaction occurred:

$2KMnO_4(s) \rightarrow K_2MnO_4(s) + MnO_2(s) + O_2(g)$

5.53 g of potassium permanganate was used in this experiment. Calculate the mass of oxygen, O_2, which forms. **(3)**

4 2.74 g of a sample of solid oxide of lead, Pb_3O_4, decomposes when heated according to the equation below.

$2Pb_3O_4(s) \rightarrow 6PbO(s) + O_2(g)$

a Calculate the mass of PbO which would be formed on heating 2.74 g of Pb_3O_4 to constant mass. **(3)**

b Calculate the volume of oxygen gas which would be produced by heating this sample to constant mass. (1 mole of any gas occupies a volume of 24 000 cm^3.) **(2)**

5 An insoluble unknown metal hydroxide has the formula $M(OH)_2$, where M represents the metal M. 3.0 g of solid $M(OH)_2$ is added to 50.0 cm^3 of 1.0 mol dm^{-3} hydrochloric acid with stirring. Some solid $M(OH)_2$ is observed lying at the bottom.

The solution is filtered and the residue washed with some water and dried. The mass of unreacted $M(OH)_2$ is 0.55 g.

$$M(OH)_2 + 2HCl \rightarrow MCl_2 + 2H_2O$$

Calculate the relative formula mass of $M(OH)_2$ and identify the metal M. **(5)**

6 a 0.2 g of calcium metal was reacted with excess nitric acid. Calculate the volume of hydrogen gas which was produced in this reaction. (1 mole of any gas occupies 24 dm^3.) **(2)**

$$Ca + 2HNO_3 \rightarrow Ca(NO_3)_2 + H_2$$

b If the concentration of the nitric acid was 2.0 mol/dm^3, calculate the volume of nitric acid needed to react completely with the 0.2 g of calcium. **(2)**

c The calcium nitrate produced in this reaction is often used in fertilisers. Calculate the percentage of nitrogen, by mass, present in calcium nitrate. **(2)**

7 To determine the concentration of a solution of barium hydroxide, $Ba(OH)_2$, 25.0 cm^3 of the solution was placed in a conical flask with a few drops of phenolphthalein indicator and titrated with a solution of hydrochloric acid of concentration 0.2 mol dm^{-3}.

$$Ba(OH)_2 + 2HCl \rightarrow BaCl_2 + 2H_2O$$

The results of the titration are shown in Table 6.3.

Table 6.3

Initial burette reading/cm³	Final burette reading/cm³	Titre/cm³
0.0	22.8	22.8
0.0	22.4	22.4
0.0	22.5	22.5

Calculate the concentration of the barium hydroxide solution in mol dm^{-3}. **(4)**

8 25.0 cm^3 of dilute battery acid was placed in a conical flask and a few drops of phenolphthalein indicator were added. Sodium carbonate solution of concentration 0.16 mol dm^{-3} was then added slowly to the conical flask until the end point was obtained. 31.25 cm^3 of sodium carbonate solution was required to neutralise the sulfuric acid in the dilute battery acid.

$$Na_2CO_3 + H_2SO_4 \rightarrow Na_2SO_4 + CO_2 + H_2O$$

Calculate the concentration of the sulfuric acid in mol dm^{-3}. **(3)**

9 Solid sodium hydrogen carbonate decomposes when heated:

$$2NaHCO_3 \rightarrow Na_2CO_3 + H_2O + CO_2$$

3.36 g of sodium hydrogen carbonate was placed in a test tube and heated in a Bunsen flame.

a Calculate the mass of sodium carbonate formed. **(3)**

b Calculate the volume of carbon dioxide produced in this reaction. (1 mole of gas occupies $24\,dm^3$ at room temperature and pressure.) **(2)**

10 Butan-1-ol was reacted with an excess of propanoic acid in the presence of a small amount of concentrated sulfuric acid.

$$C_4H_9OH + CH_3CH_2COOH \rightleftharpoons CH_3CH_2COOC_4H_9 + H_2O$$

6.0 g of the alcohol produced 7.4 g of the ester. Which one of the following is the percentage yield of the ester? **(1)**

A 57%

B 70%

C 75%

D 81%

11 Which one of the following are the units of K_c for the equilibrium, $2SO_2 + O_2 \rightleftharpoons 2SO_3$? **(1)**

A $mol^{-1}\,dm^{-3}$

B $mol\,dm^{-3}$

C $mol^{-1}\,dm^{3}$

D $mol\,dm^{3}$

12 A sample of solid phosphorus was burned in excess oxygen. 0.775 g of phosphorus reacted with 1.000 g of oxygen.

a Calculate the empirical formula of the oxide of phosphorus formed. **(3)**

b Given that the M_r of the oxide of phosphorus is 284.0, calculate the molecular formula of the oxide. **(1)**

13 Name the shape of a dichlorodifluoromethane molecule (CCl_2F_2) and the shape of a chlorine trifluoride molecule (ClF_3). **(2)**

14 Ammonia reacts with oxygen to form nitrogen(II) oxide, NO, and steam.

$$4NH_3 + 5O_2 \rightarrow 4NO + 6H_2O$$

Calculate the mass of ammonia required to react with excess oxygen to form 4.0 g of nitrogen(II) oxide given a 50% yield. **(3)**

15 Methane reacts with water as shown in the equilibrium below.

$$CH_4(g) + H_2O \rightleftharpoons CO(g) + 3H_2(g)$$

The equilibrium concentrations, in $mol\,dm^{-3}$, of each gas at a particular temperature are: CH_4, 0.14; H_2O, 0.55; CO, 0.17; H_2, 0.51.

Write an expression for the equilibrium constant, K_c, for this equilibrium and calculate its value at this temperature. **(2)**

16 A reacts with B according to the equilibrium

$$3A(g) + 2B(g) \rightleftharpoons E(g)$$

At 550 K, $K_c = 96.2\,mol^{-4}\,dm^{12}$. The equilibrium mixture contained 28.0 mol of A and 113.4 mol of B in a $140\,dm^3$ container.

Write an expression for the equilibrium constant, K_c, for this equilibrium and calculate the concentration, in $mol\,dm^{-3}$, of E in the equilibrium mixture. **(3)**

17 A burette has error $\pm 0.05\,\text{cm}^3$. In a titration, the initial burette reading was $0.05\,\text{cm}^3$ and the final burette reading was $26.35\,\text{cm}^3$. What is the percentage uncertainty in the titre value? **(2)**

A-level only questions

1 The results of a kinetic experiment are shown in Table 6.4.

Table 6.4

Experiment	Initial concentration of NO/mol dm^{-3}	Initial concentration of O$_2$/mol dm^{-3}	Initial rate/ mol dm^{-3} s^{-1}
1	4×10^{-3}	1×10^{-3}	6×10^{-4}
2	8×10^{-3}	1×10^{-3}	24×10^{-4}
3	12×10^{-3}	1×10^{-3}	54×10^{-4}
4	8×10^{-3}	2×10^{-3}	48×10^{-4}
5	12×10^{-3}	3×10^{-3}	162×10^{-4}

Using these results, write a rate equation for the reaction. Include a value for the rate constant at this temperature and deduce its units. **(6)**

2 Gases A and B react according to the following equation

$$2A(g) + B(g) \rightarrow C(g) + D(g)$$

The initial rate of reaction was measured in a series of experiments at a constant temperature.

Table 6.5

Experiment	Initial [A]/mol dm^{-3}	Initial [B]/mol dm^{-3}	Initial rate/mol dm^{-3} s^{-1}
1	1.7×10^{-2}	2.4×10^{-2}	7.40×10^{-5}
2	5.1×10^{-2}	2.4×10^{-2}	2.22×10^{-4}
3	8.5×10^{-2}	1.2×10^{-2}	9.25×10^{-5}
4	3.4×10^{-2}	4.8×10^{-2}	5.92×10^{-4}

Write a rate equation for the reaction at this temperature and calculate a value for the rate constant (k) at this temperature and deduce its units. **(5)**

3 The rate equation for the reaction of propanone with iodine in acidic solution is:

$$\text{rate} = k[CH_3COCH_3][H^+]$$

Which one of the following are the units of k? **(1)**

A $\text{mol}\,\text{dm}^3\,\text{s}^{-1}$

B $\text{mol}^{-1}\,\text{dm}^3\,\text{s}^{-1}$

C $\text{mol}^2\,\text{dm}^{-6}\,\text{s}^{-1}$

D $\text{mol}^2\,\text{dm}^6\,\text{s}^{-1}$

4 $Na_2CO_3(s) \rightarrow Na_2O(s) + CO_2(g)$

For this reaction, $\Delta H° = +323\,\text{kJ}\,\text{mol}^{-1}$ and $\Delta S° = +153.7\,\text{J}\,\text{K}^{-1}\,\text{mol}^{-1}$

Show that the thermal decomposition of sodium carbonate is not feasible at $1250\,\text{K}$.

Calculate the temperature at which the value of $\Delta G = 0$ for this reaction. **(5)**

5 The absolute entropy value for $NH_3(g) = 193\,J\,K^{-1}\,mol^{-1}$.

The boiling point of ammonia is $240\,K$.

Calculate the absolute entropy value of $NH_3(l)$.

$$NH_3(g) \rightarrow NH_3(l); \Delta H^\circ = -23.35\,kJ\,mol^{-1} \qquad \textbf{(4)}$$

Note: question 6 does not apply to CCEA candidates.

6 a Sulphuryl chloride, SO_2Cl_2, dissociates at high temperatures according to the equation: $SO_2Cl_2(g) \rightleftharpoons SO_2(g) + Cl_2(g); \Delta H = +93\,kJ\,mol^{-1}$

When 2.0 moles of sulphuryl chloride were allowed to dissociate at a given temperature, the equilibrium mixture was found to contain 1.5 moles of chlorine at a total pressure of $150\,kPa$. Calculate the value of the equilibrium constant K_p for this reaction and state its units. **(4)**

b Sulphuryl chloride is rapidly hydrolysed by water:

$$SO_2Cl_2 + 2H_2O \rightarrow H_2SO_4 + 2HCl.$$

Calculate the pH of the solution made by dissolving $132.2\,g$ of sulphuryl chloride in water to make $1\,dm^3$ of solution. **(3)**

7 Propanoic acid (CH_3CH_2COOH) has a pK_a value of 4.87. Calculate the pH of a $0.05\,M$ solution of the acid. **(3)**

8 The K_a for ethanoic acid (CH_3COOH) is $1.74 \times 10^{-5}\,mol\,dm^{-3}$. Calculate the pH of a $0.1\,mol\,dm^{-3}$ solution of ethanoic acid. Give the answer to two decimal places. **(2)**

Note: question 9 does not apply to CCEA candidates.

9 Ozone is 30% dissociated at equilibrium according to the following equation:

$$2O_3(g) \rightleftharpoons 3O_2(g)$$

Calculate the value of K_p for this dissociation reaction, stating its units, if the total pressure is 10 atmospheres. **(4)**

10 In the study of the reaction between chlorine and bromide ions:

$$Cl_2(aq) + 2Br^-(aq) \rightarrow Br_2(aq) + 2Cl^-(aq)$$

the following results were obtained.

Table 6.6

$[Br_2]/mol\,dm^{-3}$	0.00	0.14	0.21	0.28	0.34	0.37	0.37
Time/s	0	10	20	30	40	50	60

a Plot a graph of concentration of bromine against time. **(3)**

b The rate at concentration $0.100\,mol\,dm^{-3}$ is $1.17 \times 10^{-2}\,mol\,dm^{-3}\,s^{-1}$. Draw a tangent to the curve at $0.300\,mol\,dm^{-3}$ and calculate the rate at this point. **(2)**

Appendix

The table below shows the specification for maths content for AS and A-level chemistry in the required maths skill column. These skills are common to all boards. The table also illustrates where these maths skills could be developed or assessed during the course. Skills shown in **bold** type will only be tested in the full A-level course, not the AS level course. It is important that you realise that this list of examples is not exhaustive and these skills could be developed in other areas of specification content from those indicated. The information in the table is intended as a guide only. You should refer to your specification for full details of the topics you need to know.

Table 7.1

Required maths skill	Exemplification (assessment not limited to the examples below) Learners may be tested on their ability to:	Page	Is it relevant to AQA, OCR, Edexcel, WJEC/ Eduqas?	Is it relevant to CCEA?
1 Arithmetic and numerical computation				
0.0 Recognise and make use of appropriate units in calculations	▪ convert between units, e.g. cm^3 to dm^3 as part of volumetric calculations	7	✔	✔
	▪ **give units for an equilibrium constant or a rate constant**	16		
	▪ understand that different units are used in similar topic areas, so that conversions may be necessary, e.g. entropy in $J\,mol^{-1}\,K^{-1}$ and enthalpy changes in $kJ\,mol^{-1}$	14		
0.1 Recognise and use expressions in decimal and ordinary form	▪ use an appropriate number of decimal places in calculations, e.g. for pH	20	✔	✔
	▪ carry out calculations using numbers in standard and ordinary form, e.g. use of Avogadro's number	25		$pV = nRT$ not required
	▪ **understand standard form when applied to areas such as (but not limited to) K_w**	26		
	▪ convert between numbers in standard and ordinary form	25		
	▪ understand that significant figures need retaining when making conversions between standard and ordinary form, e.g. $0.0050\,mol\,dm^{-3}$ is equivalent to $5.0 \times 10^{-3}\,mol\,dm^{-3}$	36		
0.2 Use ratios, fractions and percentages	▪ calculate percentage yields	27	✔	✔
	▪ calculate the atom economy of a reaction	29		
	▪ construct and/or balance equations using ratios	29		

0.3 Make estimates of the results of calculations (without using a calculator)	▪ **evaluate the effect of changing experimental parameters on measurable values, e.g. how the value of K_c would change with temperature given different specified conditions**	33	✔	✔
0.4 Use calculators to find and use power, **exponential and logarithmic functions**	▪ carry out calculations using the Avogadro constant ▪ **carry out pH and pK_a calculations** ▪ **make appropriate mathematical approximations in buffer calculations**	26 61 59	✔	✔
2 Handling data				
1.1 Use an appropriate number of significant figures	▪ report calculations to an appropriate number of significant figures given raw data quoted to varying numbers of significant figures ▪ understand that calculated results can only be reported to the limits of the least accurate measurement	37 23	✔	✔
1.2 Find arithmetic means	▪ calculate weighted means, e.g. calculation of an atomic mass based on supplied isotopic abundances ▪ select appropriate titration data (i.e. identification of outliers) in order to calculate mean titres	42 43	✔	✔
1.3 Identify uncertainties in measurements and use simple techniques to determine uncertainty when data are combined	▪ determine uncertainty when two burette readings are used to calculate a titre value	46	✔	✔
3 Algebra				
2.1 Understand and use the symbols: =, «, », >, <, ∝, ~, ⇌	no exemplification required	50	✔	✔
2.2 Change the subject of an equation	▪ carry out structured and unstructured mole calculations ▪ **calculate a rate constant k from a rate equation** ▪ **carry out calculations using the Arrhenius equation**	50 58 57	✔	Arrhenius equation not on specification
B.2.3 Substitute numerical values into algebraic equations using appropriate units for physical quantities	▪ carry out structured and unstructured mole calculations ▪ **carry out rate calculations** ▪ **calculate the value of an equilibrium constant**	53 56 57	✔	K_p not needed, just K_c

2.4 Solve algebraic equations	• carry out Hess's law calculations	53	✔	Arrhenius equation not on specification
	• **calculate a rate constant k from a rate equation**	58		
	• **carry out calculations using the Arrhenius equation**	57		
B.2.5 Use logarithms in relation to quantities that range over several orders of magnitude	• **carry out pH and pK_a calculations**	59	✔	Arrhenius equation not on specification
	• **carry out calculations using the Arrhenius equation**	57		
4 Graphs				
3.1 Translate information between graphical, numerical and algebraic forms	• interpret and analyse spectra	45	✔	Arrhenius equation not on specification
	• **determine the order of a reaction from a graph**	73		
	• **derive rate expression from a graph**	73		
	• carry out calculations using the Arrhenius equation	57		
3.2 Plot two variables from experimental or other data	• plot concentration–time graphs from collected or supplied data and draw an appropriate best-fit curve	64	✔	✔
3.3 Determine the slope and intercept of a linear graph	• calculate the rate constant of a zero order reaction by determination of the gradient of a concentration–time graph	68	✔	Arrhenius equation not on specification
	• **carry out calculations using the Arrhenius equation**	57		
B.3.4 Calculate rate of change from a graph showing a linear relationship	• calculate the rate constant of a zero order reaction by determination of the gradient of a concentration–time graph	73	✔	✔
B.3.5 Draw and use the slope of a tangent to a curve as a measure of rate of change	• determine the order of a reaction using the initial rates method	76	✔	✔
5 Geometry and trigonometry				
4.1 Appreciate angles and shapes in regular 2D and 3D structures	• predict/identify shapes of, and bond angles in, molecules with and without a lone pair(s), e.g. NH_3, CH_4, H_2O etc.	85	✔	✔
4.2 Visualise and represent 2D and 3D forms including two dimensional representations of 3D objects	• draw different forms of isomers	85	✔	✔
	• **identify chiral centres from a 2D or 3D representation**	89		
4.3 Understand the symmetry of 2D and 3D shapes	• describe the types of stereoisomerism shown by molecules/complexes	88	✔	✔
	• **identify chiral centres from a 2D or 3D representation**	86